Antioxidants
and EXERCISE

Antioxidants
and EXERCISE

Jan Karlsson, PhD
Independent Consultant

Human Kinetics

Library of Congress Cataloging-in-Publication Data

Karlsson, Jan, 1940-
  Antioxidants and exercise / Jan Karlsson.
    p. cm.
  Includes bibliographical references and index.
  ISBN 0-87322-896-0
  1. Antioxidants.   2. Exercise--Physiological aspects.   3. Free
radicals (Chemistry)--Pathophysiology.   4. Sports medicine.
  I. Title.
    [DNLM: 1. Exercise--physiology.   2. Antioxidants--metabolism.
  3. Free Radicals--metabolsim.   QT 260 K185a 1996]
  RB170.K37   1997
  612'.044--dc20
  DNLM/DLC
  for Library of Congress   96-18843
                                                              CIP

ISBN: 0-87322-896-0

Copyright © 1997 by Jan Karlsson

Acquisitions Editor: Richard Washburn; **Developmental Editor:** Nanette Smith; **Managing Editor:** Henry Woolsey; **Editorial Assistant:** Coree Schutter; **Copyeditors:** Felice Bassuk, Holly Gilly, Dawn Roselund; **Indexer:** Theresa J. Schaefer; **Text Designer:** Judy Henderson; **Graphic Artist:** Sara Wolfsmith; **Illustrator:** Jennifer Delmotte; **Cover Designer:** Jack Davis; **Printer:** Edwards Brothers

Human Kinetics books are available at special discounts for bulk purchase. Special editions or book excerpts can also be created to specification. For details, contact the Special Sales Manager at Human Kinetics.

Printed in the United States of America   10   9   8   7   6   5   4   3   2   1

Human Kinetics
Web site: http://www.humankinetics.com/

United States: Human Kinetics, P.O. Box 5076, Champaign, IL 61825-5076
1-800-747-4457
e-mail: humank@hkusa.com

Canada: Human Kinetics, Box 24040, Windsor, ON N8Y 4Y9
1-800-465-7301 (in Canada only)
e-mail: humank@hkcanada.com

Europe: Human Kinetics, P.O. Box IW14, Leeds LS16 6TR, United Kingdom
(44) 1132 781708
e-mail: humank@hkeurope.com

Australia: Human Kinetics, 57A Price Avenue, Lower Mitcham, South Australia 5062
(08) 277 1555
e-mail: humank@hkaustralia.com

New Zealand: Human Kinetics, P.O. Box 105-231, Auckland 1
(09) 523 3462
e-mail: humank@hknewz.com

To my "grown ups," sailing and skiing companions—Anne, Eva, Thomas, and Nallo (a golden retriever).

Contents

I

Introduction to Nutraology

This section elaborates on the scholarship of nutraology in sports medicine and how radical and antioxidant biology were first introduced to explain muscle injury. The intent of the book is to present antioxidant vitamins as nutraceuticals; the additional roles of vitamins (for example, as coenzymes) are also described to distinguish them from the antioxidant function.

Some food supplement programs have already been established in sports medicine. In most cases, such programs have advantaged sport performance; however, diet manipulation has risks. The significance of a nutritious, well-balanced diet as the basis for successful nutratherapy must not be minimized.

1

Introduction

Nutraology is a new medical field that has developed in the last decade, with nutratherapy as its clinical application. Nutraology is based in such disparate fields as internal medicine, cardiology, oncology, physiology, and nutrition. For some 10 years, scientists in these fields have been studying a new metabolic entity: radical formation and the biology of antioxidants.

History of Radical and Antioxidant Science

In the mid- and late 1970s, my students and I investigated muscle histochemical lesions in otherwise healthy adults, endurance athletes, and patients with intermittent claudication. A few years later, we began to include patients with cardiac failure in our research program. We decided to seek cause-and-effect relationships between the lesions and muscle metabolism.

Until that time the prevailing concept had been that a major determinant in muscle exercise performance was central circulation, measured by such means as maximal oxygen uptake and cardiac output. The periphery, to which muscle mass belonged, was only "a sink for oxygen." The term *respiration,* in its physiological context, was originally confined to pulmonary functions—not to respiration of sugar and fat in some cell organelles referred to as *mitochondria.* Later on, even mitochondrial metabolic turnover was found to be involved in regulation of central circulation.

The (re-)introduction of the muscle biopsy needle, and the subsequent muscle electrolyte, metabolic, and histochemical studies slowly altered that picture, bringing the periphery more and more attention. Thus, maximal oxygen uptake became confined to the distribution of the oxygen using slow-twitch muscle fibers and distribution of the peripheral blood flow.

These new methods also allowed us to study muscle injuries, which were subsequently found to be muscle-fiber related in elite athletes, fitness sport participants, and patients with cardiac disorders. Runners frequently experience a crippling pain in their lower leg muscles, referred to as *runner's foot*, which also appears to have these muscle-fiber-related characteristics [Wallensten and Karlsson 1984a; Wallensten and Karlsson 1984b]. Other names for these symptoms are *soldier's foot, shin splints, truck-driver's foot* (in Australia), and the more scientific term *closed-compartment syndrome.*

Until the 1980s, muscle injury had been studied in relation to high-energy phosphate metabolism, lactate accumulation, and pH decrease [Hillar and Schwartz 1972; Opie 1965]. Most scientists believed that lactic acid formation and subsequent changes could explain histochemical lesions. But prolonged exercise without any apparent lactate formation was also found to cause muscle trauma [Gollnick, Ianuzzo, and King 1971].

A common denominator for these histochemical lesions was mitochondrial swelling. This feature was investigated as early as the mid-1960s by a group of molecular cardiologists based at the University of Göteborg in Sweden [Ekholm et al. 1968]. In their studies they emphasized membrane-bound phospholipids. Ischemia was found to deplete the mitochondria on structural fats (e.g., lysolecithin), which affects the metabolism of polyunsaturated fatty acids (PUFA), in general, and essential fatty acids (EFA, or vitamin F), in particular. Fat as a fuel, on the other hand, increased in these experimental situations [Gudbjarnason et al. 1968].

In summary, back in the early 1980s, muscle histochemical lesions to structural properties could be discussed as the result of both pH and other signs of failure in electron transport and/or handling.

During my postdoctoral training in the early 1970s in the Department of Internal Medicine at Southwestern Medical School in Dallas, Texas (now a branch of The University of Texas), my roommate was Giorgio P. Littarru, now professor of cardiology at the Catholic University in Rome, Italy. Collaborating with Dr. Karl Folkers, from the Institute for Biomedical Research at The University of Texas at Austin, he studied the significance of ubiquinone (coenzyme Q) in cardiac failure [Littarru, Ho, and Folkers 1972]. We had frequent discussions about muscle metabolism and muscle failure. As a result, I gained a unique insight into the cellular action of ubiquinone.

When I returned to Stockholm and had my first doctoral student to advise, one of his PhD dissertation committee members, the renowned biochemist Dr. Lars Ernster (now a retired professor of biochemistry at the University of Stockholm but also a past president of the International Biochemistry Society), spoke about the relation between lactate accumulation and subsequent trauma. He advised us to look into muscle metabolism and the significance of ubiquinone. In a series of papers he had written, he had already addressed the role played by ubiquinone in mitochondria and the handling of electrons. He emphasized that ubiquinone "had a vitamin E-like effect"; at that time vitamin E was recognized as a protective agent that, although there was not precise knowledge of its biochemical action, could be linked to electron transfer [Lehninger 1965]. Thus, my research group and I had a solid scientific basis from which to begin to investigate possible approaches to the study of muscle trauma based on Dr. Lars Ernster's initiative.

The concept of radical and antioxidant biology in medicine dates back to the early 1980s, when it was suggested that the biological activity of antioxidants was due to their inhibiting lipid peroxidation in biological membranes by scavenging chain-propagating radicals [Burton and Ingold 1981; Burton et al. 1983]. Two decades earlier, however, biochemists and biophysicists had already started to reveal the biology of radicals, the formation of "active oxygen species," and their scavenging [Michaelis 1946; Chance 1947].

The role of one investigator, Dr. Britton Chance, must be mentioned. Thanks to the work that he and his associates did, our knowledge of molecular oxygen metabolism, the formation of oxygen radical species, and the role of ubiquinone has been expanded, providing from the start an integrated interpretation [Chance, Schoener, and Chindler 1964; Chance and Pring 1968]. Chance is not only a sailing companion but also my mentor in advanced biochemistry and biophysics. In addition to Dr. Chance, such scholars as Dr. Martin Klingenberg and Dr. Peter Mitchell must be cited for their contributions in the field of biochemistry. Dr. Mitchell won a Nobel Prize for his research on the mitochondrial membrane, in which ubiquinone was given a central role.

In the early 1980s I began to collaborate with Dr. Karl Folkers, and for a decade we published many articles jointly. At first Dr. Folkers was not inclined to appreciate the antioxidative properties of ubiquinone, or coenzyme Q_{10} as he preferred to call the compound. He was more fond of its function in mitochondria as a coenzyme. I am not sure that he ever will fully accept ubiquinone as an antioxidant. We did agree, however, that ubiquinone was a significant nutrient, and in our joint work we gave the compound the name vitamin Q.

The scholarly validity of the use of the term *vitamin Q* has been challenged, especially by nutritionists. But by then, I had already read my pre-med textbook in biochemistry, *Introduction to Modern Biochemistry*, in which ubiquinone was referred to as vitamin Q (table XIX, p. 379) [Karlson, 1965]. The same interpretation has also been put forward by Geigy [Geigy 1986], as described in table 2.2.

Topics to Be Discussed

This book summarizes the knowledge that we have acquired so far and reviews it in relation to the findings of other investigators. The main topics to be discussed follow:

- Radical formation in healthy muscle tissue, which takes place almost exclusively in the lipoidic layers where most of the respiratory enzymes are located
- The control of radical formation at this particular level by fat-soluble (lipophilic) antioxidants to avoid deteriorating peroxidation
- The elucidation of not only the lipophilic antioxidants vitamins Q and E but also the compounds (such as EFA, or vitamin F) that are most susceptible to peroxidation and at the same time lipophilic in nature [Karlson 1965]
- The "thresholds" in the total antioxidant activity: (1) vitamin Q to catalyze the quantitatively most important lipophilic antioxidant—vitamin E, and (2) vitamin C to unload vitamin E with the water-soluble (hydrophilic) antioxidants (Referred to as the *vitamin Q-E-C cycle*.)
- The exploitation of the hydrophilic antioxidants as vitamins C and P, such as bioflavonoids or the phenol-ring-based compounds present in most vegetables and fruits, or in products based on these sources
- The primarily cytosol-based antioxidant enzymes, which return the electrons to the respiratory chain and retrieve oxygen or other radical modified molecules
- The biology of the peroxidation-sensitive vitamin F, in general, and the subgroup vitamin F_1 (omega-3 fatty acids), in particular
- The links between these biochemical reactions and processes and cell trauma to the muscle, red and white blood cells, and extracellular compounds
- Muscle exercise, radical formation, and antioxidant activity
- The nature of the antioxidant mechanisms and chain-breaking processes and their relationship to endogenous processes and to food intake, nutrition, and food quality

Applicability of Nutratherapy to Recreational Exercise and Elite Sport

Based on an analysis of the scientific background, the applicability of nutratherapy to recreational exercise and elite sport is presented and discussed with respect to these matters:

- The diet alone and food supplement programs (nutratherapy)—a matter of food quality
- The theory of scientifically based nutratherapy programs
- Muscle exercise before and after nutratherapy
- Performance gains with nutratherapy on the leisure sport and elite athlete levels
- Ethical aspects of nutratherapy in relation to food additives

My reason for writing this book is to present this topic as objectively and precisely as possible based on present knowledge. It is clearly not my ambition to present all scientific arguments for or against nutratherapy as a clinical intervention in medicine or in sport. In my opinion, that matter has already been satisfactorily settled. I will, however, engage in the debate of the ethical aspects of nutratherapy in sport. A number of people—scholars, coaches, and sport administrators—have expressed their opinions in this matter in such a way that many participants in fitness sport and elite athletics have interpreted nutratherapy as being in "an ethical gray zone" between nutrition and pharmacological intervention programs. This has tempted some nutritionists and fitness advocates to compare nutratherapy to doping as it is presently defined by international sport authorities.

2

Historical Perspective

This chapter elaborates on the scholarship of nutraology in medicine and how radical and antioxidant biology were first introduced to explain the background of and later to protect the athlete from muscle injury. Many antioxidants were previously known as nutrients or vitamins. Other, more recently discovered nutrients with antioxidative properties (e.g., vitamins Q and P) are not recognized as vitamins by all scholars and scientific disciplines. They are described as vitamin-like substances or as "unofficial or obsolete" vitamins.

This chapter introduces definitions of the terms applicable to the rest of the book: *food supplements, nutraceuticals*, which are based on nutrients, and *food additives*. The latter group consists of "cosmetic additions" to our diet such as caffeine from coffee, ginsenosides from ginseng root, and ethanol from wine. The rationale for using the term *vitamin* for all these essential or conditionally essential nutrients is also given.

Although the intent of the book is to present antioxidant vitamins as nutraceuticals, the additional roles of vitamins (for example, as coenzymes) are also described to distinguish them from the antioxidant function.

Finally, it should be noted that some food supplement programs have already been established in sports medicine. In most cases, these programs have proven to be advantageous in sport performance, but there are risks with diet manipulation. The significance of a nutritious, well-balanced, mixed diet as the basis for a successful nutratherapy must be emphasized.

In recent years, many scientific papers have been published whose contents overlap the fields of nutrition, physiology, biochemistry, and biophysics, and that address the topics of the biology of radical formation and antioxidant mechanisms. Exercise physiologists and orthopedic surgeons recognized muscle histochemical lesions in the early 1980s as a consequence, or even a cause, of overuse injuries (see figure 2.1). In 1986 the American College of Sports Medicine (ACSM) and Dr. Robert R. Jenkins arranged a symposium [Jenkins 1986] in which the biological and pathological backgrounds of radical-induced injury were presented. The potential of this research for sports medicine was also outlined by the speakers.

Introduction of New Concepts

The International Society for Myochemistry was organized in the early 1980s by Drs. Noris Siliprandi (one of the inventors of synthetic L-carnitine), Giorgio Benzi, and Lester Packer. Representatives from different scholarly fields were invited to meet and discuss joint issues. The society organized meetings and published proceedings in the fields of muscle metabolism, in general, and radical biochemistry, in particular.

At that time, two major points were proffered:

- A muscle adaptation existed to improve the antioxidant capacity [Jenkins et al. 1984] (see figure 2.2a).
- Intervention programs were suggested by one of the clinical researchers in the field, Dr. Harry Demopoulus [Demopoulos et al. 1984] and were proven to reduce radical-induced muscle injury [Packer and Viguie 1989] (see figure 2.2b).

The major components of these intervention programs were *nutrients*. A nutrient is a food constituent (e.g., water, vitamin, mineral, protein, fat, and carbohydrate) upon which people have developed a biological-nutritious dependence. Most of the nutrients used in these programs were vitamins, such as vitamins E and A [NRC 1989a]. These nutrients were administered in concentrated forms as *food supplement products (nutraceuticals)*. Thus, a nutraceutical is an industrially produced entity based on either a food remedy or an artificially synthesized chemical molecularly identical to a food remedy. A nutraceutical is applied in a nutratherapy program according to present knowledge of nutraology.

Nutraology is part of the scholarly fields of physiology and nutrition, covering research in

- nutrient needs to prevent disease,
- design and safety testing of nutraceuticals,
- safe application of nutraceuticals in intervention programs in health care and preventive medicine, and
- distinguishing physiological effects of nutraceuticals from potential pharmacological and toxic properties of the same molecular entities.

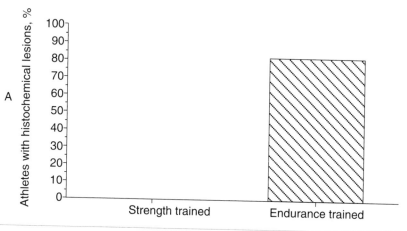

Figure 2.1a Sport epidemiological studies on muscle histochemical lesions in strength- and endurance-trained elite athletes [Sjöström, Johansson, and Lorentzon 1987].

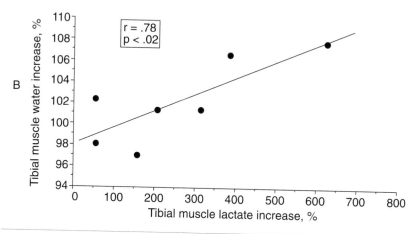

Figure 2.1b Closed compartment syndrome, muscle lactate increase, and muscle water increases [Wallensten and Karlsson, 1984b; Wallensten and Karlsson, 1984c].

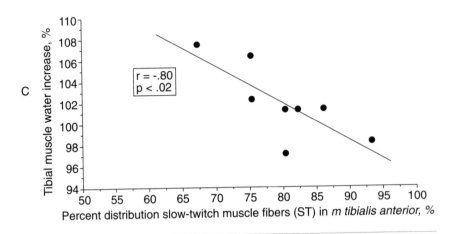

Figure 2.1c Closed compartment syndrome, muscle lactate increase, and muscle quality expressed as percent distribution of slow-twitch fibers (%ST) [Wallensten and Karlsson, 1984b; Wallensten and Karlsson, 1984c].

Over the last 3,000 to 6,000 years, people have become accustomed to "cosmetic" additions to nutrient intake, such as caffeine from coffee, ginsenosides from ginseng root, and ethanol from wine and other alcoholic beverages. These cosmetic additions are referred to as *food additives*. Other ingredients assimilated in our diet might be a part of a certain molecular structure in the body (e.g., cystein in glutathione). The extent to which these ingredients represent nutrients or precursors of an endogenous synthesis might be regarded as a matter of semantics. This question, however, has sparked a discussion among biochemists, nutritionists, and pharmacologists. Intervention programs using these ingredients could be seen as supplying food supplements—nutratherapy or nutraceutical therapy—or as pharmaceutical intervention. It is plausible that an extra intake of, for example, branched-chain amino acids (BCAA) could be argued to represent a pharmaceutical rather than a nutraceutical therapy.

Nutratherapy and Health Promotion

Food has been used as an elixir since antiquity. According to the Eber Scrolls, the ancient Egyptians used liver to treat night blindness. Hippocrates introduced the term *diaita* to describe the significance of food intake in health and disease. Our word *diet* derives from that idea.

This particular field of nutrition—therapy or prophylactic measures—is also a concern in modern times. New terms have been introduced by

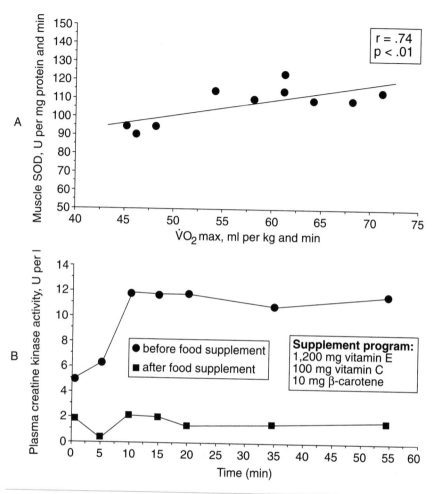

Figure 2.2   (a) The individual relationship between muscle superoxide dismutase (SOD) activity in *m vastus lateralis* and pulmonary maximal oxygen uptake [Jenkins, Friedland, and Howald 1984]. (b) The beneficial effects of a food supplement program based on lipophilic antioxidants with regard to prevention of muscle overuse injury [Packer and Viguie 1989].

scientists, research organizations, the pharmaceutical industry, and their corporate research and marketing departments, such as

- nutraceutical,
- designer food, and
- functional food.

Thus, the distinction between food and pharmaceuticals is sometimes blurred. The existence of concentrated nutrients—

nutraceuticals—either as industrial, artificial, or synthetic products, or as industrially refined, biologically synthesized products has put pressure on the regulatory authorities of industrialized countries, including Japan. Matters of concern for the industry are clinical trials, labeling, safety of substances, approval routines, regulatory classification, and so on.

Terminology is essential in all scientific contexts, and the author has adopted the following terms:

food supplement—Nutrients in separate and concentrated forms that supplement normal dietary nutrients. Also known as a *nutraceutical.*

food additives—Food "extras," where a defined nutritional dependence is so far unproven and the substances can be referred to as *food cosmetics*

nutraceutical therapy—nutratherapy—Food supplements administered as therapy and prophylactically

Food supplements, food additives, and nutraceutical therapies are based on nutrients in concentrated form and sometimes as artificial products and are therefore comparable to an ethical drug therapy.

Table 2.1   The internationally recognized list of nutrients referred to as vitamins [NRC 1989 a and b].

Vitamin name	Generic name	Solubility
vitamin A	retinol	fat
vitamin B_1	thiamine	water
vitamin B_2	riboflavin	water
vitamin B_3	niacin	water
vitamin B_5	pantothenic acid	water
vitamin B_6	pyridoxine	water
vitamin B_{12}	cobalamin	water
vitamin C	ascorbate	water
vitamin D	calciferol	fat
vitamin E	α-tocopherol	fat
vitamin K	menaquinone	fat

Table 2.2   Some Nutrients Recognized as "Unofficial" Vitamins

"Vitamin" name	Generic name	Solubility
Vitamin F_1*	Omega-3 fatty acid	Fat/water
Vitamin F_2*	Omega-6 fatty acid	Fat/water
Vitamin H (alt B_x)**	Biotin	Water
Vitamin B_y**	Pteridine	Water
Vitamin J*	Choline	Water
Vitamin M	Folic acid	Water
Vitamin O (B_{11})	Carnitine	Fat/water
Vitamin P (C_2)	Bioflavonoids (e.g., rutin)	Water
Vitamin Q**	Ubiquinone	Fat

* Vitamins recognized by the Food and Drug Administration (FDA 1979). The terms F_1 and F_2 have been proposed by the author to distinguish between omega-3 and omega-6 fatty acids.
** Vitamins recognized by *Geigy Scientific Tables* 1986.

Vitamin Terminology

Another matter of concern is the definition of the term *vitamin*. Eleven nutrients (see table 2.1) are accepted by the World Health Organization (WHO) as true vitamins.

Other nutrients, for reasons unknown to the author, have not been accepted or have been excluded from the vitamin list. Some of these nutrients are listed in table 2.2.

An example of recent research results bearing on food intake, nutrients, and antioxidant properties are the (bio-)flavonoids, also referred to as *vitamin P* [Kühnau 1976; Geigy and Geigy 1962]. The flavonoids have disappointed all early therapeutic hopes. However, in a recent population study (the Zutphen Elderly Study), their protective effect on cardiovascular diseases was described [Hertog et al. 1993]. These nutrients are today heralded as explaining the epidemiological phenomenon referred to as the *French paradox* or the *Mediterranean diet*. Formerly, beta-(ß-) carotene was exclusively recognized as "pro-vitamin A." Today it is discussed, among other things, as a significant antioxidant that protects plasma LDL from peroxidation [Cabrini et al. 1986; Zamora,

Table 2.3   Radicals and Their Targets in Uncontrolled Reactions

Vitamin F_1 or Ω-3	Vitamin F_2 or Ω-6
Alpha-(α)linolenic (ALA)*	Linoleic acid (LA)*
Eicosapentenoic acid (EPA)	Gamma- (γ-)linolenic
Docosapentenoic acids (DPA)	Dihomogamma-(g-)linolenic acid (DGLA)
Docosahexenoic acid (DHA)	Arachidonic acid (AA)

* As nutrients, the first representative of each series is the true essential fatty acid (EFA). Because the others in each column are synthesized under conditions that do not guarantee ample amounts of each respective acid to be produced in all situations, they are referred to as conditionally essential fatty acids.

Hidalgo, and Tappel 1991]. Vitamin A, on the other hand, is cited less and less often as an antioxidant but rather as a signal substance involved in cell mitosis regulation. Vitamin A and its 10 to 15 molecule varieties act as intra- and extracellular signal substances with hormone-like properties. In this context, similarities exist between vitamin A and another lipophilic but nonantioxidant vitamin—vitamin D.

Geigy Scientific Tables does not define the term *vitamin*. It lists, however, fat- and water-soluble vitamins in vol. 4, page 61 [Geigy 1986]. *Geigy Scientific Tables* include as vitamins, in addition to those listed in table 2.1, ubiquinone (vitamin Q), pteridine (vitamin B_y), and biotin (vitamin H or B_x). (See table 2.2.)

The essential fatty acids (EFA) were the sixth group of relatively homogeneous nutrients that have "essential properties" (see table 2.3). In the 1920s they were given the letter F, because A through E were already assigned to other essential nutrients. The term *vitamin F* has since been applied on and off in the scientific literature [Geigy and Geigy 1962; Karlson 1965].

Obviously, then, the terminology applied to vitamin-like nutrients is not consistent. This inconsistency has caused intense debate through the decades among scholars of different fields. It is not my ambition to evoke old interdisciplinary debates and conflicts or to excavate buried and forgotten hard feelings. It is my intention, however, to evaluate the pros and cons of both camps based on our present knowledge, and through such a process improve understanding of the subject.

The term *vitamin* is scientifically a "black box," having no clear definition or meaning. The rationale for using the term in describing nutrients in general is obvious to those who do research and lecture in this field. The word *vitamin* means to both scholars and the general population a food constituent to which humans have developed, over millions of years, a certain nutritious relationship or requirement, whether or not its biochemistry is known. Knowledge, whether complete or not, must be made available to physicians and laypeople so that it can be communicated to both fitness sport participants and elite athletes. The term *vitamin* has in that respect a clear pedagogical advantage and will enhance information communication.

That vitamins are constituents in our diet goes without saying. Too low an intake could cause a shortage and lead to symptoms of a deficiency disease.

Table 2.4   Enzymes and Their Cofactors

I.   For a complete biological action:

          enzyme protein + cofactor(s)

II.   Cofactor:

    a. coenzyme

      1. cosubstrate (e.g., NAD)

      2. prosthetic group (e.g., FAD)

    b. inorganic ion

III.   Holoenzyme = apoenzyme + cofactors

IV.   Apoenzyme = only the enzyme protein

Vitamins and Enzymology

Since the mid-1980s, much interest has been focused on vitamin Q, or ubiquinone [Karlsson 1987]. Vitamin Q, a newly discovered nutrient, was found in relation to the cell's organelle, the mitochondrion, where cellular respiration takes place. Respiration is equivalent to combustion, but is stepwise controlled, whereas combustion is a free radical auto-oxidation process [Sawyer 1988]. Vitamin Q was found to have a specific action in these stepwise reduction oxidation (red-ox) reactions and electron transport. Because of that particular function, the newly discovered compound was named *coenzyme Q* [Crane et al. 1957; Lester and Crane 1959].

The term *coenzyme* is confined to the action of proteins (see table 2.4). Most enzymes consist of a protein (the apoenzyme entity) with the additional component of *cofactors*. The cofactors can be either organic compounds of low molecular mass (or *coenzymes*) or inorganic ions. Taken together, they are referred to as *holoenzymes*. As to the coenzymes, they are differentiated as either a *cosubstrate* (e.g., a hydrogen donor) or a *prosthetic group*. A prosthetic group is more tightly bound to the enzyme protein (e.g., flavine adenine dinucleotide, or FAD).

Table 2.5   Compounds in Our Food Sources That Have Been Attributed
Antioxidant and Disease-Fighting Properties

Compound	Food Source	Function
Allylic sulfides	Garlic and onion	Glutathione precursor
Carotenoids	Carrots, parsley, vegetables	Antioxidant, vitamin A precursor
Bioflavonoids (catechins, tannins)	Tea, red wine, vegetables, fruits	Antioxidants
Indoles	Cabbage, brussels sprouts	Block steroid hormone synthesis
Thiocyanates	Horseradish, radish	Detoxification
Limonoids	Citrus fruits	Detoxification
Lycopens	Tomatoes	Antioxidants
Monoterpenes	Vegetables	Antioxidants

Later research demonstrated that only a fraction (about 30%) of all vitamin Q takes part in its role as a coenzyme. The rest (around 70%) is active as an unspecific antioxidant in most cells and tissues and their organelles [Beyer, Nordenbrand, and Ernster 1986; Kalén et al. 1987; Crane 1990]. Vitamin Q's role both in formation of radicals and in the cell's protection against radicals—the antioxidant function—is well recognized today [Ernster et al. 1992].

The development of the field of nutrients and their biological significance has been amazing. Even in the subfields, the amount of literature published is enormous. Table 2.5 presents some of the organic compounds in different food sources that are currently attracting the attention of biologists, pharmacologists, pharmacists, and the pharmaceutical industry as potential antioxidant compounds or nutrients.

Earlier Nutraceutical Concepts

In the late 1960s, as a junior researcher and doctoral candidate, I participated in the muscle glycogen loading research programs carried out jointly by different departments at Karolinska Institute in Stockholm, Sweden. With the (re-)introduction of the needle muscle biopsy technique, it became possible to describe the biology of an earlier observation made by Dr. Erik Hohwü Christensen, Chairman of the Department of Exercise Physiology. The well-known concepts in sports medicine of a carbohydrate-enriched diet and muscle glycogen loading were reasserted [Saltin 1973]. In addition to their scholarly contributions, these research efforts were recognized for their contributions to sports medicine and the well-being of endurance athletes [Karlsson and Saltin 1971]. But other sports, such as ice hockey [Åkermark et al. 1996] and soccer [Jacobs et al. 1982] have also benefitted from the knowledge as it was applied in sport practice.

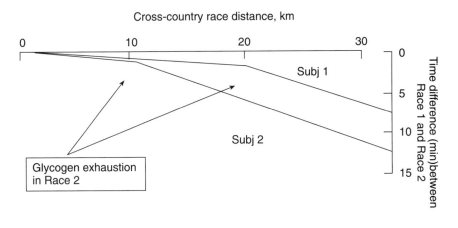

Race 1: after glycogen supercompensation
Race 2: after mixed diet ("normal stores")

Figure 2.3   The beneficial effects of a carbohydrate supplement program on muscle glycogen loading ("supercompensation") and the subsequent physical performance in a cross-country running race by two volunteers [Karlsson and Saltin 1971].

Earlier observation on the effects of carbohydrate supplementation was done in the 1930s. Under the guidance of Nobel Prize laureate Dr. August Krogh, at Copenhagen University in Denmark, Dr. Christensen had observed how those who adhered to a carbohydrate-enriched diet for a couple of days improved endurance exercise performance [Christensen 1939a, b, c].

Dr. Bengt Saltin (one of my mentors) and his group described how the food supplement program, based on a carbohydrate-enriched diet, enhanced deposition of carbohydrates as muscle glycogen in healthy people. Moreover, Dr. Christensen's original observation concerning endurance performance in sports could be confirmed (see figure 2.3) [Karlsson and Saltin 1971; Saltin 1973]. Earlier, another group in Stockholm, guided by Dr. Jonas Bergström, had shown the significance of muscle glycogen loading on exercise performance under experimental laboratory conditions [Bergström and Hultman 1966].

Nutratherapy as a Potential Risk Factor

The carbohydrate-enriched diet and the consequent muscle glycogen loading have been widely accepted since the late 1960s as an important way to prepare for endurance sports and training [Karlsson, Nordesjö, and Saltin 1974]. The dietary program, however, was meant to be applied only occasionally. But since both competition-related endurance

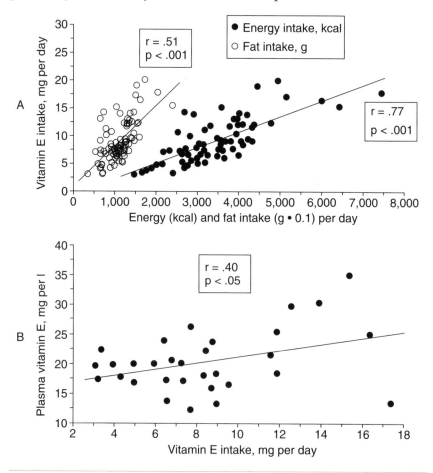

Figure 2.4   (a and b) The individual relationship between the estimated vitamin E intake and the food and fat intake for 1 week, according to questionnaires filled out by elite male and female cross-country skiers (Karlsson, in press).

and training endurance were improved with this dietary program, it was developed into a long-term treatment program and was used not only by elite cross-country skiers and long-distance runners but also by professional tennis players. In a recently released book from the International Olympic Committee (IOC) Medical Commission, it was suggested that the carbohydrate (CHO) intake as a fraction of the total energy intake would correspond to 60% to 65% (E%CHO) [Ekblom 1994]. The corresponding figures for lipids (E%L) and protein (E%Pr) were 15% (E%L) and 20% to 25% (E%Pr), respectively.

Such long-term dietary regimens are synonymous with malnutrition. It has been shown that the intake of lipophilic nutrients such as vitamin E is linearly related to fat intake (see figure 2.4) [Karlsson, Diamant, Edlund et al. 1992; Karlsson, Diamant, Theorell et al. 1993]. Individuals can take as little as 8 to 10 mg a day of vitamin E. Even if that figure corresponds to the recommended daily allowance (RDA) of this particular nutrient according to the National Research Council (NRC) [NRC, 1989a], it represents only a minute fraction of what, for example, the United States Olympic Committee (USOC) recently deemed safe for nutraceutical therapy in elite athletes (i.e., 400 mg a day) [Grandjean 1994].

Other risks are associated with such an extreme diet that is followed for a long time. Providing a maximal oxygen uptake ($\dot{V}O_2$max) corresponding to 6 L per minute in, for example, a male cross-country skier and 3 hours of training a day, it could be computed that approximately 120 g of fat are combusted. Assuming a caloric intake corresponding to 8,000 kcal (34 megajoules) a day and an E%L of 15%, the lipid intake

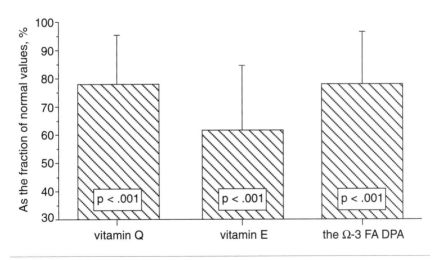

Figure 2.5   The plasma values for the antioxidant vitamins Q and E and the essential fatty acid, DPA, of the omega-3 series in cross-country skiers expressed as a percent of the normal values of sedentary healthy persons (Karlsson, in press).

would be in the vicinity of 130 mg a day, which would not allow any net turnover of, for example, the EFA (or vitamin F): the omega-3 (Ω-3) and omega-6 (Ω-6) fatty acid series. In fact, this dietary regimen means that occasionally structural lipids, which to a large extent are polyunsaturated fatty acids, will be recruited for fuel purposes.

This metabolic situation will induce deficits with respect to lipophilic nutrients such as vitamins Q and E and PUFA, in general, and vitamins F_1 and F_2 (EFA), in particular (see figure 2.5) [Karlsson, Diamant, & Theorell et al. 1993]. Vitamins Q and E are significant for white blood cells—they are the cells richest in antioxidants—and consequently for the immune system [Bendich 1993]. Essential fatty acids are important to maintain necessary fluidizing properties of different membrane systems (e.g., cell walls) [Ghosh, Dick, and Ali 1993].

Radical Trauma in People Who Exercise

The survival of red blood cells (RBC) is related to their cell wall omega-3 fatty acid content. Low omega-3 fatty acid content will cause a state of physical stiffness, increased susceptibility to impact physical trauma (fragility) in, for example, the capillary passage, and the possibility of hemolysis [Beving, Petrén, and Vesterberg 1990]. Most likely, RBC fragility is an additional explanation for runners' anemia besides the well-documented plasma expansion with endurance training [Packer 1986; Saltin et al. 1968]. The WBC also depend on omega-3 fatty acids for their plasticity, which is an essential entity for migration and capillary wall penetration as a result of activation, migration, chemotaxis, and immune function [Berg Schmidt et al. 1991].

Athletes with an extremely high intake of carbohydrates and subsequent impaired intake of lipid-based or lipophilic nutrients have been in a situation referred to as the *carbohydrate syndrome*, or the *carbohydrate trap* [Karlsson, Diamant, Edlund et al. 1992; Karlsson, Diamant, Theorell et al. 1993b]. It seems reasonable to assume that this condition might reduce resistance to radical formation, bio- and histochemical lesions, cell injury, and inflammatory processes, referred to as *overuse injuries* in athletes [Packer 1986]. Overuse injuries might, in turn, reduce physical properties in muscle and connective tissues and subsequently increase the risk of impact injury [Jones, Cowan, and Knapik 1994]. Recent research has shown that impaired glucose uptake in the contracting muscle could be related to local radical trauma in elite endurance athletes [Karlsson and Rønneberg 1996]. The same kind of muscle lesion is absent in moderately physically active men [Karlsson, Lin, Sylvén, and Jansson 1996].

Radical trauma to skeletal muscle has also been documented in patients with ischemic cardiovascular disease [Karlsson, Lin, Gunnes, et al. 1996; Karlsson and Semb 1996]. It has been debated to what extent these findings represent adaptation processes or are comparable to the kind of injuries seen in elite endurance athletes [Karlsson and Rønneberg 1996].

There is an overwhelming amount of literature available today describing radical biochemistry and the potential threat of radical species to the living cell. Radical formation and radical injury are also present in healthy humans as a normal biological feature. The role of certain nutrients as antioxidants has also been accepted in general terms. As to the prophylactic and therapeutic use of antioxidants, there is much more conjecture and fiction than facts where muscle activity, increased metabolic rate, and oxidative stress are concerned [Jenkins 1993].

Summary

A good diet is a prerequisite for high performance by elite and recreational athletes. The carbohydrate-enriched diet, in which carbohydrates and carbohydrate supplements fortify the diet and boost muscle glycogen, is an example of a high-performing diet.

Food sources contain nutrients that are essential in the cell's defense mechanisms against reactive radical species. Many of these nutrients have been known for decades and have been recognized as vitamins (e.g., vitamin E). Others are known only through more recent research efforts: As no new vitamins have been accepted since 1952, they are only unofficially recognized as vitamins or vitamin-like nutrients. Some examples are ubiquinone (vitamin Q) and bioflavonoids (vitamin P).

Physical conditioning and increased respiratory enzyme activities in contracting muscles cause increased antioxidant enzyme activities and allocation of antioxidant nutrients to these muscles. Nutraceutical therapy programs with these nutrients and vitamins have also been proven to increase the antioxidant capacity and possibly, by indirect means, physical performance.

II

Radical Formation

This section introduces the scientific background of radical and antioxidant biology and its applicability to sports medicine. Chapter 3 discusses the biochemistry of radical formation as it occurs in the exercising muscle. The ways in which the antioxidant system are attached, designed, and adapted to fit different aspects of radical formation and subsequent reactions are described. Radical formation is a "spillover" of the respiratory metabolism in the muscle mitochondria. The reader will see that radical formation is not only a threat; radicals as semimanufactures are exploited in cell regeneration, but they are also the backbone of our immune system.

Chapter 4 further outlines the antioxidant system and establishes the basis for the theory of nutratherapy. The phenol-ring molecular structure of many antioxidant vitamins as a common denominator is discussed. Fat solubility (lipophilicity) and water solubility (hydrophilicity) are presented as dependent on added chains and other structures and as constituting a basis for the antioxidant strategy.

Mitochondrial, microsomal, and other organelle membranes and their lipids host the radical-producing enzymes. The role of the lipophilic antioxidants—vitamins Q and E—is described.The ways in which the antioxidant potential is expanded by vitamin C's catalyzing property in the lipid-water interface between vitamin E and vitamin C are discussed. Taken altogether, they constitute the vitamin Q-E-C cycle. The nutratherapy concept outlined later in the book is based on this relationship. This critical sequence makes the remaining hydrophilic antioxidant systems available: the remaining vitamin C pool, vitamin P (bioflavonoids), glutathione, antioxidant enzymes, and other nutrients and metabolites with antioxidant properties.

Also discussed is the issue of whether the catalyzing property of vitamin Q, in addition to the fact that vitamin Q keeps the other antioxidant vitamins in a necessary chemically reduced state, bears on vitamin Q's endogenous quality. Vitamins E and C, in contrast, are exclusively exogenous nutrients.

Chapter 5 reviews basic stoichiometric chemistry and how it is applicable to (re-)cycling systems in biology. Recycling systems are necessary in the action of antioxidant enzymes and their oscillation between reduced and oxidized states. As a model, vitamin Q is used as a coenzyme in the mitochondria and their electron shuttling. This, then, bears on the mechanisms by which vitamin Q as a coenzyme keeps the remaining cellular content of vitamin Q (around 70%) and other antioxidant vitamins in a chemically reduced state and active as antioxidants.

Chapter 5 also discusses the "scavenger-like" reaction of polyunsaturated fatty acids (PUFA) on an irreversible reaction. As an example, peroxidation-sensitive PUFA are used. After peroxidation, they are further metabolized and excreted or expired as gases. The fate of other cell constituents such as hyaluronic acid or the genome (DNA and RNA) in a radical-rich milieu is also discussed as an impaired muscle mechanical function and a potentially reduced supercompensation after a training session. Chapter 5 also includes a section on the antioxidant enzymes that depend on muscle quality indicated by muscle fiber composition and training level.

Chapter 6 applies the knowledge about radical biochemistry to living tissue, in general, and to exercising muscle, in particular. In light of the extremely tight control of oxygen availability in the resting and exercising muscle, the "normal" oxygen radical formation corresponding to 5% to 15% of oxygen turnover is discussed. The accumulation of metabolites confined to the fermentation process (anaerobiosis) and their local as well as central, ergoreceptor-mediated stimulation are explained. Downregulation of the nitric oxide (NO·)-dependent endothelium derived relaxing factor (EDRF) by oxygen radical species as an alternative or additional mechanism of capillary dilation and in regulation of peripheral resistance is discussed.

The larger allocation of antioxidant vitamins to muscle tissue or fibers plus the higher antioxidant enzyme activities in muscle tissue or fibers adapted to a high oxygen turnover is an intriguing feature. This is discussed in light of animal exercise studies with and without food supplements.

The radical formation of WBC and their attack on foreign material is a backbone in our immune defense system. This situation is discussed in relation to inflammatory processes in muscle and joints as WBC, paradoxically, play a significant role in the overuse injury syndrome.

Chapters 7 and 8 further apply this information to exercising muscles. Radical trauma to lipids, proteins, the genome, and so on, although controversial, does exist, and antioxidant vitamins and enzymes are available to control and protect. Through the food chain we get all these essential or conditionally essential nutrients. The food sources are exposed to the same radical threat as humans, and vegetable and animal products have to obey the same biological rules as the human body does for their protection. Polyunsaturated fatty acids are as susceptible to peroxidation in the salmon swimming in the North Atlantic and Pacific oceans as they are in the human heart. The biological consequences of this are spelled out.

Is a well-balanced, mixed diet always appropriate with respect to nutrient intake? In chapter 9, this question is addressed with the focus on sports medicine. Muscle exercise per se takes its toll on antioxidant nutrients, and different modes of exercise modify this toll. In light of this, direct and indirect support for the radical trauma concept is advanced. The discussion also questions whether it is satisfactory to include only antioxidants in nutratherapy, or if one of the major targets for radical trauma—vitamin F_1 (omega-3 fatty acids)—should also be included. The issue of whether macro- and microminerals are needed is also addressed.

Lipophilic nutratherapy ingredients, such as triglycerides containing omega-3 fatty acids, can be hydrolyzed before and absorbed by the intestinal mucosa as water-soluble acids. Other lipophilic nutrients such as vitamins Q, E, and A and the "pro-vitamin A" (beta-carotene) demand lipids, which can form micelles (microscopic lipid droplets). The lipophilic nutrients are dissolved in the micelles and taken up as an entity by the mucosa. The relevance of that to intestinal absorption, lipid contents in the diet, and plasma lipids is commented upon in chapter 10. This might also be relevant to why LDL cholesterol in cardiology is frequently referred to as the "bad cholesterol fraction"—the vitamin E content in LDL particles could be depleted.

3

Principles of Radical Formation

Electrons and electron exchange are the basis for reduction-oxidation (red-ox) reactions. The electron-accepting compound will then be reduced and the donor oxidized. Electrons are also essential in the process of radical species formation.

The term *radical* belongs to the same family of chemical and physical terms as atom, molecule, ion, and electron. The atom, or for that matter the molecule (i.e., a stable assembly of atoms interconnected with covalent electron bindings), has no charges. The number of positive particles (protons) equals particles with a negative charge (electrons). Ions have either positive or negative charges. The number of charges equals the number of lost or added electrons to the atom or molecule.

Definition of the Term *Radical*

A *radical* is an entity with one or several *unpaired* electrons in the outer electron orbit of the molecule. The entity can be an atom or a molecule, charged or uncharged. The unpaired electron is usually extremely exchangeable, which is the chemical and physical reason for the reactivity of the radical species. This is also the radical property on which the scavenging activity of antioxidants is based [Lopez 1990].

Most radicals are extremely unstable, energy-rich entities resulting from unpaired electrons. Electron exchange is why radicals are so quick to participate in chemical reactions. These chemical reactions enable the radical species to decrease its energy level and its reactivity potential. Nature is opposed to concentration—even of energy. In most cases, a reactive radical species will by this reaction form a less reactive radical species. The final product can be an extremely stable radical (i.e., a harmless species), but it can also have enzymatic reactions whereby the electron is recovered by normal red-ox processes and, with respect to cellular metabolism, forms water.

The level of reactivity will determine the life span or survival of the radical species. Most radicals do exist only during a fraction of a second before they have participated in a chemical reaction. Some radicals have obtained an extremely low energy level and are therefore referred to as *stable radicals* [Sawyer 1988].

Radicals can be uncharged or have negative or positive charges. The property that singles out the radical is its unpaired electron in the outermost electron circuit (orbit). An uncharged radical is a molecule in which one of the electrons is excited and translocated to its own separate circuit. This is true for the radical nitric oxide (NO·), which sometimes is referred to as the body's own "endogenous nitrate." NO· is significant in capillary dilation, which is of utmost importance to enhance not only oxygen delivery to the muscle but also in, for example, lactate efflux.

A positive or a negative radical (sometimes incorrectly referred to as a *radical ion*) has either lost or obtained a single electron. The remaining unpaired electron in the outer orbit explains the reactivity of the radical species. Depending on the major atom(s) of the species, the radical is referred to as

- a carbon (C)-centered radical,
- a nitrogen (N)-centered radical,
- a sulfur (S)-centered radical,
- an oxygen (O)-centered radical, and so on.

NO· is an example of a nitrogen-centered radical.

Metabolism and Radical Formation

Molecular oxygen or dioxygen (O_2) is a gas present in the air at approximately 24% volume. When the first living organism appeared on the earth some 4 to 5 billion years ago, oxygen was only present in chemical bindings or oxides. The existing organisms based their metabolism on fermentation of energy-rich compounds for their energy needs. They are therefore referred to as *anaerobic* organisms. Some of them, or at least their relatives, might even exist today.

The next major step in the evolution of life on earth occurred some 3 to 4 billion years ago, when photosynthesis applied by marine bacteria for energy transformation and survival was introduced, with radiation energy in the form of sunlight as the energy source. These primitive forms of bacteria used hydrogen sulfide (H_2S) as the source of hydrogen atoms as reducing equivalents in their red-ox processes. With the development of blue-green algae some 1 billion years later, water (H_2O) replaced hydrogen sulfide as a source of reducing equivalents. Molecular oxygen was formed as a by-product.

The molecular oxygen formed was at first used to precipitate iron oxide in the earth's oceans. At that time, atmospheric oxygen concentration was maintained, according to estimates, at levels less than 1%. When this necessary titration of ocean iron was completed, there was an almost instant increase in atmospheric oxygen to 17% to 21% [Sawyer 1988].

Until then, life was restricted to the aquatic environment of the ocean. The atmosphere contained no ozone to protect life from the solar ultraviolet radiation, which has enough energy to destroy complicated molecules such as organic compounds.

The appearance of physically dissolved molecular oxygen in the oceans killed most of the existing fermentative organisms. Some species adapted and maintained their fermentation metabolism (anaerobes); others retreated to an oxygen-free environment and survived there; still others evolved to oxygen-using aerobic organisms (aerobes). Cellular respiration was evoked. Combustion is a free radical auto-oxidation process, whereas respiration is a controlled, stepwise breakdown process of fuels [Sawyer 1988].

The terms *aerobic* and *anaerobic* are among the most misused terms in biology and medicine. The origin of the terms dates back to the 1920s and 1930s and the early days of cellular biology and medicine and corresponding theories. They should be replaced with such terms as *respiration* and *fermentation*.

A prerequisite for both respiration and combustion is molecular oxygen, which is eager to adopt electrons from its surroundings. A typical electron donor is atomic or molecular hydrogen (H_2). In the mitochondria of the cell and in the process line referred to as the *respiratory chain*, cellular respiration leads in controlled stepwise reactions to formation of water. Hydrogen originates from the breakdown of carbohydrates and fat, whereas oxygen is transported out by the circulatory system.

As a result of blue-green algae and their photosynthesis, molecular oxygen appeared in the atmosphere as ozone (O_3). This was a prerequisite for life to be able to leave the marine environment and to safely establish a terrestrial life protected by the radiation shield as offered by the atmospheric ozone layer. All terrestrial life bears a memory of their marine origin. Salt water is the evolutionary origin of extracellular fluids such as plasma or embryonic water in multicellular organisms, including humans.

The transition from photosynthesis plus fermentation to respiration plus fermentation by these primitive organisms has provoked many speculative theories as to the evolutionary steps involved, including speculation about the invasion of extraterrestrial, primitive aerobic organisms. To make a long story short, respiration of organic molecules with a photosynthesis background and transformation of the chemically bound energy demanded the introduction of metalloproteins for an effective use of dioxygen as an oxidant.

Molecular oxygen has the ability to accept or receive one or several electrons—e^-—from its immediate surroundings (see figure 3.1). These electrons might be free or attached to a more or less stable molecule or ion. Electron transfer in mitochondrial processes is loosely bound to transport vehicles such as the cofactors NAD, NADP, FAD, and enzymes (see table 2.4). These electrons can easily be accessed or annexed by molecular oxygen.

POTENTIAL MOLECULAR OXYGEN (O_2) REACTIONS

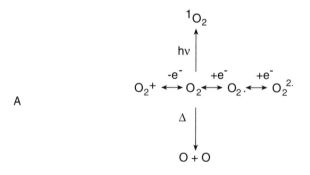

Figure 3.1a   The reactions in which molecular oxygen (O_2) could possibly participate. Dissociation of O_2 in atomic oxygen (O + O) is inconceivable in a biological environment as the energy level (about 4-5,000°K) is never attained. Singlet oxygen ($^1O_2$), however, is produced as a result of both mechanical (vibration) or radiation energy conduction and metabolically derived energy.

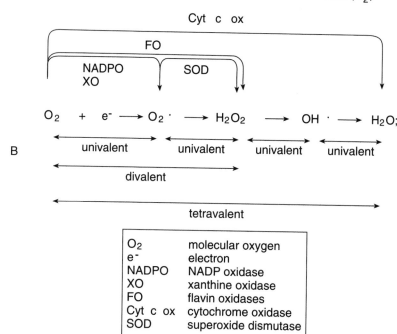

Figure 3.1b   Molecular oxygen can participate in univalent, divalent, and tetravalent reactions in the mitochondrial metabolism. Respiration and water formation is the normal and tetravalent reaction. Water can also be formed by four consecutive univalent reations, which means oxygen-centered radical formation and hydrogen peroxide (H_2O_2) as well. Hydrogen peroxide will provoke radical (OH⁻) formation from water.

C
$$\overset{+}{\cdot}\overset{-}{N}\text{-}O \longleftrightarrow \cdot N\text{=}O \longleftrightarrow \overset{-}{N}\text{=}\overset{+}{O}. \longleftrightarrow N\text{=}O \cdot$$

Figure 3.1c   The nitric oxide molecule, which is a gas, exists in different molecular constellations in resonance with each other.

Superoxide and Hydroxyl Radicals

Even if reactions with molecular oxygen are the basis for the energy-releasing respiration processes, oxygen is almost never free, but is bound to closely related proteins: hemoglobin in the red blood cells and myoglobin in heart and skeletal muscle—hence, the red color of blood and muscle tissue. If free oxygen will come close to an electron in an exchange phase or collide with a free electron, the superoxide radical—$O_2^{\cdot-}$—or *active oxygen* will be formed in a univalent reaction (see figure 3.1).

$$O_2 + e^- = O_2^{\cdot-}$$

The superoxide radical is an example of an oxygen-centered radical (oxyradical). The superoxide radical is constantly formed in the mitochondria at a rate determined by the actual mitochondrial oxygen turnover. It is estimated that, under certain conditions, up to 15% of molecular oxygen in mammals might go through one or more oxygen-centered radical species stages [Sawyer 1988]. During muscular exercise, the corresponding figure is estimated at 3% to 5% [Demopoulos et al. 1984].

This is quite a substantial radical formation in exercising humans. Pulmonary oxygen uptake is in the order of 2 liters per min (L × min^{-1}) in sedentary people and 5 to 7 liters per min in endurance athletes of both genders. These figures correspond to 1 to 3 µmol O_2 per mg muscle per min (µmol × mg^{-1} × min^{-1}). For fitness sport participants and elite athletes, producing this free oxygen radical source in their contracting muscles is suggested to be the major reason for radical-related cell and tissue trauma, or *overuse trauma* [Davies et al. 1982].

The superoxide radical can (and most probably will) react with a water molecule. Water is present as fluid (e.g., in the cytosol or in plasma) or as water molecules physically attached to proteins, lipids, and carbohydrates in the "water-free" lipid layers, or in any kind of intracellular membrane or the cellular wall. Under these conditions, the more reactive hydroxyl radical species (OH$^{\cdot}$) is formed:

$$H_2O + 1/2\ O_2^{\cdot-} = 2OH^{\cdot}$$

The hydroxyl radical must be distinguished from the stable anion: the hydroxide ion—OH$^-$. This ion is present everywhere water is present and is harmless under conditions of a normal pH.

Hydrogen Peroxide and "Singlet" Oxygen

Oxygen can also participate in divalent and tetravalent reactions (see figure 3.1b). The latter reaction forms water (H_2O), which is one of the end products in mitochondrial respiration. Mitochondria can also undertake divalent reactions, where hydrogen peroxide (H_2O_2) is formed. Hydrogen peroxide is not a radical but can participate in reactions that produce radicals. In that respect, it is a pro-oxidant.

The superoxide radical can spontaneously or enzymatically dismutase and form hydrogen peroxide.

$$O_2^{\cdot-} + 2\ H^+ \rightarrow H_2O_2$$

Another oxygen species that is sometimes referred to as a radical is "singlet" oxygen. Molecular oxygen is a relatively stable molecule with six electrons ("paired electrons") in the outer orbit. This molecule can be energetically excited, which can occur as a result of a neighboring chemical reaction where energy is released. Energy can also be exposed on the oxygen molecule from external sources such as intense mechanical vibration or radiation (gamma [γ] radiation, ultraviolet [UV] radiation in sunlight, etc). As a result, one of the six electrons in the outer orbit will then jump into a new "extra" outer orbit. Under these conditions, molecular oxygen can achieve an unpaired electron, which is in accordance with the definition of a radical.

The energized state of molecular oxygen in the form of singlet oxygen has some similarities with the molecule nitric oxide (NO·) (see fig. 3.1c). Nitric oxide has recently met with great appreciation in biology and medicine because it is a signal substance in capillary dilation [Bassenge 1992]. In its biological context, nitric oxide is mostly referred to as a radical species. But, as is true for molecular oxygen, it exists also in biology in resonance with other molecular forms.

The energized stage of molecular oxygen and transfer of this energy package to cellular molecular structures can be deteriorating and harm the cell. Antioxidants are therefore used to quench singlet oxygen.

Perhydroxyl Radical

In the presence of hydrogen protons, the superoxide radical can form the perhydroxyl radical (HO_2·). This radical species is a moderately strong radical that can oxidize allylic carbons and thereby act as an initiator of cell trauma.

Radical Formation in the Service of Life

Radical formation is a normal biological process necessary in cellular synthesis processes, including formation of DNA and RNA and certain hormones (see figure 3.2). These reactions engage molecular oxygen, hydrogen peroxide, and specific substrates catalyzed by a vast array of metalloproteins with one or more transition metals (iron [Fe], copper [Cu], manganese [Mn], or molybdenum [Mb]).

The "deadly threat" that radicals represent to biological life is also exploited by white blood cells in their immune and aseptic activities. Superoxide and other radicals are formed by the outer section of the cell wall of activated white blood cells. These radicals will attack foreign material as virus particles, bacteria, fungi, and

RADICAL FORMATION IN THE BIOLOGICAL MILIEU I

RADICAL FORMATION MECHANISMS IN THE CELL:

1. enzymatic processes

   in energy metabolism—combustion in mitochondria

   a. "spillover" — a normal biological feature
   b. uncontrolled formation—biochemical lesions

   by xanthine oxidase (XO)

   XDH → XO (initiated by leukocytes)

   by lipoxygenases; Ω-6 fatty acid arachidonic acid

   in peroxisomes

   catecholamines metabolism

   phagocytosis by neutrophil leukocytes

2. nonenzymatic processes (see figure 3.3)

Figure 3.2   Radical formation in the cell can be derived by means of metabolism and by enzymatic and nonenzymatic reactions (see also figure 3.3). The enzymatic processes could be normal or pathological. XDH = xanthine dehydrogenase.

other microbes weaken them so that they will be more easily accessible for phagocytosis and their ultimate destruction. These immune activities also include foreign biological material such as transplanted organs or tissues. These reactions are referred to as rejection reactions and can be moderated by immune-suppressing pharmaceuticals such as cyclosporin.

The strategy exploited to destroy foreign biological material is also used in the process of healing traumatized organs or tissues: White blood cells migrate to the area, get activated, and start bombarding the injured cells with free radicals to speed up deterioration, evacuation of breakdown products, and recovery. Certain enzymes and hormones or signal substances will appear. Due to the increased osmotic pressure, water is withdrawn from plasma and edema develops. An inflammatory process has been established. To many, these processes are associated with activation of pain receptors, discomfort, and crippling conditions for performance in both fitness and elite sport. Disorders such as rheumatoid arthritis, muscular dystrophy, atopic eczema, and psoriasis have also been linked to radical processes, although the etiology is debated [Luft 1994].

Radical Formation and Cell Protection: The Antioxidant Strategy

Reactive radical species, such as the hydroxyl radical, have to be controlled and protected from reaching cellular constituents such as lipids, proteins, and other complicated molecules including unsaturated fatty acids, hyaluronic acid, DNA, and RNA. These molecules or cell structures will immediately react with the radical and be more or less permanently rearranged, fragmented, or depolymerized. From a biological point of view, they are then in many cases permanently and irreversibly destroyed (see table 3.1).

Under normal conditions, free radical formation in the cell is actively kept under control by the antioxidant defense system—the antioxidant strategy. The majority of free radicals are generated in lipid layers, as they host the enzymes necessary to catalyze the radical-producing reactions. Major lipophilic antioxidants are vitamin Q, vitamin E, and beta-carotene. Together these lipophilic antioxidant vitamins constitute the first defense line. A later defense line includes the water-soluble vitamin C, several members of the vitamin B group, vitamin P (bioflavonoids), metabolites as urea and bilirubin, and antioxidant enzyme systems, such as the enzymes glutathione peroxidase (GPX) and reductase (GRD), their coenzymes (glutathione), or prosthetic groups (selenium, Se).

Taken together, the antioxidant strategy of the cell is based on the fact that, with few exceptions, the major radical formation starts in the lipoidic layers or cell membranes and is then transferred to the aqueous compartment(s). The lipid-soluble antioxidants (the lipophilic antioxidants) and the water-soluble antioxidants (the hydrophilic antioxidants) make up the theory of the Q-E-C cycle, with the vitamins Q and C as critical constituents with catalyzing properties. With this de-

Table 3.1 Radicals and Their Targets in Uncontrolled Reactions

Radical species	Targets	Products
$HO_2^\cdot$, $OH^\cdot$	Lipids (PUFA)	Lipid peroxides, aldehydes (e.g., MDA)
$HO_2^\cdot$, $OH^\cdot$	Proteins	Cross linking
$HO_2^\cdot$, $OH^\cdot$	Hyaluronic acid	Glycosides
$OH^\cdot$	DNA/RNA	Strand breaks, 8-hydroxyguanosine

sign, the antioxidant strategy is able to exploit not only the quantitative antioxidant properties of the vitamin E pool and the joint pools of vitamins C and P, but also the glutathione system and other antioxidant enzyme systems.

To describe these functions, the terminology has been expanded with terms such as *scavenging*, *quenching*, and *trapping*. The corresponding words (*scavenger, quencher*) are used to represent the antioxidants.

Principles of the Cascade Reaction

The hydroxyl radical species is potentially the most dangerous free radical species. If the radical formation process is not quenched and subsequently terminated, not later than at this level, essential parts of the cell's skeleton or biological machinery will be hit. The skeleton of the cell is made up of a mixture of structural lipids, proteins, and carbohydrates of the mucopolysaccharide type.

Initial and Chain Reaction

The consequences of the existence of the hydroxyl radical on the cell, provided that the antioxidant defense has failed, are as follows:

1. The hydroxyl radical hits one organic compound, represented by the symbol RH—either a lipid, a protein, or a nucleotide—and the following reaction is initiated (initial reaction):

$$RH + OH^. \rightarrow R^. + H_2O$$

   A peroxidized molecule in the form of a carbon-centered radical has been created.

2. The traumatized molecule and reactive radical can react with molecular oxygen in a chain reaction:

$$HR^- + O_2 \rightarrow ROO^.$$

   A peroxyl radical has been formed. The peroxyl radical is also an example of a carbon-centered radical.

3. The peroxyl radical reacts with a "healthy" or original lipid or protein molecule or constituent in a second chain reaction:

$$ROO^- + RH \rightarrow R^. + ROOH$$

   A hydroperoxyl radical has been formed.

If the reaction chain had started with a fatty acid as R, a fatty acid or alkyl hydroperoxide radical would have been formed. This is frequently reported as LOOH, where L stands for lipid. A lipid hydroperoxyl radical resulting from lipid peroxidation will then further deteriorate, and malondialdehyde (MDA) and/or the alkane gases pentane or ethane will be formed. Concentration determinations of plasma malondialdehyde (MDA) is a clinical tool to diagnose the presence of radical formation or inflammatory processes and estimate its magnitude. Expired pentane and ethane are also measures of radical formation.

Steps 2 and 3 in the cascade reaction depend on and create a new oxygen-traumatized molecule or radical R·, which immediately starts a novel sequence of reactions of its own. A chain reaction has been formed.

Branch Reactions

4. Iron is normally present in a chelated, harmless form, but after the trauma it can be released. Iron and other transition metals possess potentials as catalysts of reactions such as

$$2 \ ROOH \xrightarrow{Me} RO· + ROO· + H_2O$$

In Step 4, an alkyl hydroperoxide lipid molecule has created two different radicals—a branch reaction has been initiated. It is self-explanatory why this is a critical feature as two new chain reactions are formed. The combination of branch and chain reactions are the principles of the cascade reaction.

Chain reactions, as exemplified in Steps 2 and 3, will continue as long as

a. Molecular oxygen is present and
b. No antioxidant is present to break the "production line" and terminate the chain reactions.

A lipid or an alternative hydroperoxide (ROOH or RO_2H) can be substituted by another peroxyl metabolite from the mitochondrial activity, such as hydrogen peroxide (H_2O_2).

Fenton Reactions

As described earlier, red-ox reactions mean that electrons have been exchanged. Some compounds with electrons less affiliated with their molecule or ion are more easily oxidized than others; electrons can be

RADICAL FORMATION IN THE BIOLOGICAL MILIEU II

RADICAL FORMATION MECHANISMS IN THE CELL:

1. enzymatic processes (see figure 3.2)

2. nonenzymatic processes

   a. Fenton reactions

   $$Me^{n+} + O_2^{\cdot} \rightarrow Me^{(n-1)+} + O_2;$$

   $$Me^{(n-1)+} + H_2O_2 \rightarrow Me^{n+} + OH^{\cdot} + OH^-$$

   $$O_2^{\cdot} + H_2O_2 \rightarrow O_2 + OH^{\cdot} + OH^-$$

   e.g., Fe, Cu, Hg, Pb, etc.

   b. "auto-oxidation" of semiquinones

      1. endogenous (e.g., ubiquinone)
      2. exogenous (e.g., from anthracyclines)

Figure 3.3   Nonenzymatic, radical-forming processes. A major source for free radical formation is the so-called Fenton reactions, in which transition metals (Me^{n+}; $Me^{(n-1)}$)participate. Another source for radical formation is the spontaneous reactions with semiquinone from deranged mitochondria or as side effects of anthracycline drugs (e.g., Adriamycin™).

released more easily. Hydrogen is such an example and can be explosive in a certain mixing relation to oxygen (an experiment well known to most chemistry students). Hydrogen is referred to as a pro-oxidant. Other elements acting as transition metals and eager to release electrons include mercury (Hg), lead (Pb), cadmium (Cm), copper (Cu), and certain iron (Fe) ions. These pro-oxidants are also known as *heavy metals* in the terminology of industrial pollution. Other pollutants, frequently with organic backgrounds including active or passive exposure to tobacco smoke, also act as biological pro-oxidants.

The metal element iron (Fe) is necessary in the formation of vital proteins such as hemoglobin and myoglobin. Iron can form free radicals by means of the so-called Fenton reaction (see figure 3.3). Virtually all iron in the body is well stored or bound (chelated), which limits this possibility drastically. With cell trauma (e.g., muscle fiber ruptures, hemolysis, etc.), Fe can be released and lead to subsequent radical formation.

RADICALS AND REACTIVE OXYGEN SPECIES (ROS)

NAME:	FORMULA:	RADICAL:	ROS:
superoxide radical	$O_2\cdot$	+	+
singlet oxygen	$^1O_2\cdot$	+(?)	+
hydrogen peroxide	H_2O_2		+
hydroxyl radical	$OH\cdot$	+	+
alkyl radical	$R\cdot$	+	+
(alkyl-)peroxyl radical	$ROO\cdot$		+
(alkyl) hydroperoxide	$ROOH$		+
nitric oxide	$NO\cdot$	+(?)	
semiquinone (from vitamin Q)	$Q\cdot$	+	
phenoxyl radical (from vitamin E)	$E\text{-}O\cdot$	+	

"R" is an abbreviation for organic molecules in general, which frequently is substituted with "L" when exclusively lipids are considered.

Figure 3.4   Some of the best-known radicals in biology and medicine.

Biologically Significant Radicals

Progress has been made during recent years in relating specific diseases to oxidation in biology and medicine. Alzheimer's disease, atherosclerosis, diabetes, cancer, emphysema, iron overload, malaria, muscular dystrophy, Parkinson's disease, rheumatoid arthritis, and retinal degeneration are a few such examples [Scott 1995].

Frequently, the term *reactive oxygen species* (ROS) is synonymous with the term *radical.* This is obviously true only for the oxygen-centered radicals. Biochemists and biophysicists with a more fundamental approach to science have complicated the picture further by arguing that singlet oxygen ($^1O_2$) and nitric oxide ($NO\cdot$) do not represent true radicals. This quibbling is more understandable in respect to hydrogen peroxide (H_2O_2). Figure 3.4 lists some of the more significant biological radicals or ROS.

Summary

A fraction of the mitochondrial molecular oxygen (O_2) turnover will pass through one or more stages of these "active oxygen" species.

- The superoxide radical $O_2^{.-}$
- The perhydroxyl radical $H_2O^{.}$
- Hydrogen peroxide H_2O_2
- The perhydroxyl radical $HO_2^{.}$
- The hydroxyl radical $OH^{.}$

If one of the oxygen-centered radicals—the superoxide or the hydroxide radical—is able to escape the antioxidant defense line, it might hit a lipid, a protein, or a nucleotide. A peroxidized, carbon-centered, reactive radical species $(R^{.})$ has been formed in the initial reaction.

This radical could react with oxygen, and a peroxyl radical $(ROO^{.})$ is formed. This could react with a "healthy sister molecule" and a hydroperoxyl radical $(ROOH$ or $RO_2H)$ has been formed. A chain reaction has been established.

Two such hydroperoxyl radicals can react and produce two new reactive radicals, which will initiate their own chain reactions. A branch reaction is performed.

All these reactions, taken together, can be compared to the domino principle, and the term *cascade reactions* describes the events. A cascade reaction represents a serious threat to the living cell or organism. If it is allowed to continue, it is devastating, and an inflammatory process has developed. However, the inflammatory process will activate different decomposing systems including aggressive hormones, which will speed up tissue healing processes and a general recovery from symptoms.

Free radical production can occur within the framework of metabolism and enzymatic processes and can also originate from nonenzymatic reactions. An example of the latter is spontaneous superoxide radical generation from semiquinone, which could be the result of deranged mitochondria and malfunction. Transition metals and Fenton reactions provide other nonenzymatic and pathological sources of free radical formation.

4

Principles of Radical Quenching

It is not only a necessity to stop the reactive free radicals (such as the hydroxyl or peroxyl radicals) from engaging in uncontrolled reactions; it is a prerequisite for cell survival [Ernster 1986].

The evolution of life contains a number of major steps forward, or breakthroughs. As mentioned in chapter 3, cellular respiration is such a breakthrough, and a missing link for evolution biologists. The question is whether antioxidant reactions involving carbon-, nitrogen-, sulfur-, and oxygen-centered radicals could be the origin of and basis for exploitation of proton donors such as hydrogen sulfide (H_2S) and water (H_2O). Could they be the introduction of photosynthesis?

Antioxidant reactions were certainly present in the primitive forms of marine life. With metabolic adaptations to the radical threat, and for its own survival, the cell came to possess unique biochemical mechanisms. These qualities could be the origin and basis for exploitation of energy-rich compounds and later molecular oxygen for the purpose of respiration. This is a major step in the evolution of life on earth.

Uniqueness of Phenol Structures

A molecule with a relatively strong reductant property is needed to quench or scavenge (or neutralize) the unpaired electron from a free radical species. Such compounds are abundantly present in all living organisms. A common denominator for many of them is that their molecular structure is based on a carbon-ring structure, which originates from substituted phenols, polyphenol catechols, quinols, and hydroxyquinols (phenol species). These phenol species come from molecules such as hydroxylated benzene, tyrosine, or phenylalanine.

A phenol species is, consequently, an organic compound with an aromatic nucleus based on the benzene ring with one or more hydroxyl groups (hydroxygroups, -OH). The phenolic OH group is always attached to a double bond of the benzene ring, which offers a vacuum-like condition with respect to electrons. Because of this condition, this section of the phenol structure has unique properties to handle electrons.

This entity—the phenol species and its hydroxyl group(s)—is frequently referred to as the *phenol ring, group,* or *nucleus.* Biochemists differentiate between the pure antioxidant function, which is related to the nucleus, and the proton-donating feature in quenching lipid radicals, which is governed by the hydroxyl groups [Serbinova et al. 1993].

The quenching process can be exemplified by quinol (phenolate) (see figure 4.1a). This compound can release two protons to form the quinol anion and, subsequently, reduce another compound in a red-ox process. The quinol anion can then, in two separate reactions, release two electrons and form the stable entity quinone (see figure 4.1a). This molecule variant can be reduced by the same reaction steps according to the principles of reversible reactions. Consequently, it oscillates between these two extremes—quinone and quinol—providing a correct biochemical environment in regard to temperature, osmolarity, and electron flow. A driving force to maintain quinols reduced and as potential antioxidants is the electron-producing mitochondria.

The quinol entity has been demonstrated to have antioxidant properties. The fact that the electron-release reactions take place in two steps means, by definition, that two unpaired electrons (two free radicals) can get their unpaired electrons paired. The radical has subsequently been trapped, quenched, and scavenged. It also means that one intermediate (metabolite) from quinol possesses only one unpaired electron—that quinol has formed a free radical, semiquinone (the quinol radical) (see figure 4.1a). It has been suggested that the radical semiquinone may be a source of reactive radical species and trauma [Demopoulos et al. 1984]. That possibility, however, has been disputed [Beyer and Ernster 1990; Crane et al. 1991b].

In biology and medicine, many phenol-based compounds are referred to as flavonoids (bioflavonoids), flavonols, phytochemicals, or vitamin P (see table 2.2). Theoretically, they all have this reductant property. But this property varies from one phenol species to another. The chemical features granting these capacities depend on environmental factors such as temperature, pH, osmolarity, electron pressure or flow, etc. For example, the living cell operates within a relatively narrow range of temperatures (36 to 41°C) and pH (6.5 to 7.5). These narrow ranges allow only a minute fraction of these phenol compounds to act as antioxidants in a living cell. The remaining phenol compounds might, however, constitute precursors for synthesis of endogenous compounds that fit these biological specifications. Vitamin P is an intriguing feature of our antioxidant defense and explains such phenomena as the Mediterranean diet [BCCPSG 1994] and the French paradox [Keys, Aravanis, and Blackburn 1967; Taylor, Dawsey, and Albanes 1990].

The basic molecular structure could be either a mono- or polyphenol compound.The phenolic hydroxyl groups may also be chemically varied in different ways by means of methoxylation or glycosylation. The most frequent phenol-based compound in nature is said to be ubiquinone, a name derived from the Latin *ubiquitous* [Morton et al. 1957].

Ubiquinones have a phenol ring, which is a substituted phenol species (benzoquinones) with a side carbon chain originating from the mevalonate pathway with 6 to 10 isoprene units (see figure 4.1b). This carbon chain is also referred to as the *isoprene* or *farnesyl group* ("tail"). The ubiquinone moiety with 10 isoprene units, which is synthesized in most cells and tissues in humans, is designated as *ubiquinone-10* or *coenzyme Q_{10} (UQ_{10}* or *CoQ_{10})*. The rationale for using the term *coenzyme* was presented in chapter 2.

Benzene-based structures of the phenol species are synthesized in plants and microorganisms. Animal organisms do not have the necessary enzymes. One benzene-ring source for man is the essential amino acid phenylalanine. Phenylalanine, like most aromatic compounds, is formed by plants and microorganisms. Within the framework of different food chains, they accumulate in different animal protein sources. As a result, animal protein is the most efficient source of these essential compounds for man.

The phenol-ring structure has the antioxidant properties of the molecule, whereas the carbon chain (the tail) is significant for transport and retention of the molecule within membranes [Crane et al. 1991a]. In these respects, the vitamin Q molecule has almost identical properties of the vitamin E molecule [Burton and Ingold 1993] (see figure 4.1c).

Ubiquinone is present in many food sources and is available and absorbed as a nutrient. Under certain conditions, with an unsatisfactory endogenous synthesis rate, it can ultimately become a conditionally essential nutrient. This is the case when, for example, the mevalonate pathway activity is reduced [Bélichard, Pruneau, and Zhiri 1993; Folkers et al. 1990]. This has been the reasoning behind referring to ubiquinone as a conditionally essential nutrient and as vitamin Q.

In the mid-1960s, biochemistry textbooks had already grouped quinones together with vitamins and other nutrients [Karlson 1965]. At the same time, in a major Swedish encyclopedia, the term *Q-vitamin* was used for ubiquinone. CIBA-GEIGY lists ubiquinone as vitamin Q [Geigy 1986]. However, international nutrition authorities have not recognized ubiquinone as a vitamin [NRC 1989a].

Many vitamins, such as E, A, beta-carotene, and K (phylloquinone), are based on phenol structures. All of them participate in red-ox reactions and some (e.g., beta-carotene) act as scavengers including quenching singlet oxygen. In these cases, the antioxidant active part of the molecule is also the phenol ring or phenol-ring-like structures, whereas the carbon chain determines transport and membrane retention [Burton

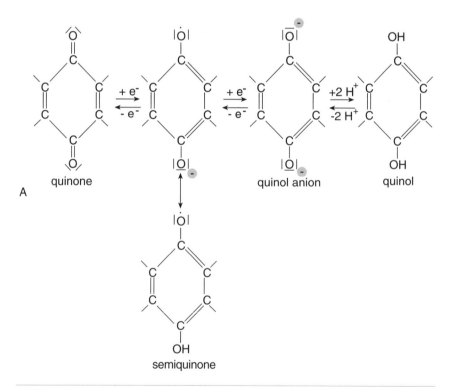

Figure 4.1a The reduction-oxidation (red-ox) reactions, where quinols are stepwise oxidized to quinones and reduced back to quinols. This is an example of sequences of reversible or cycling reactions.

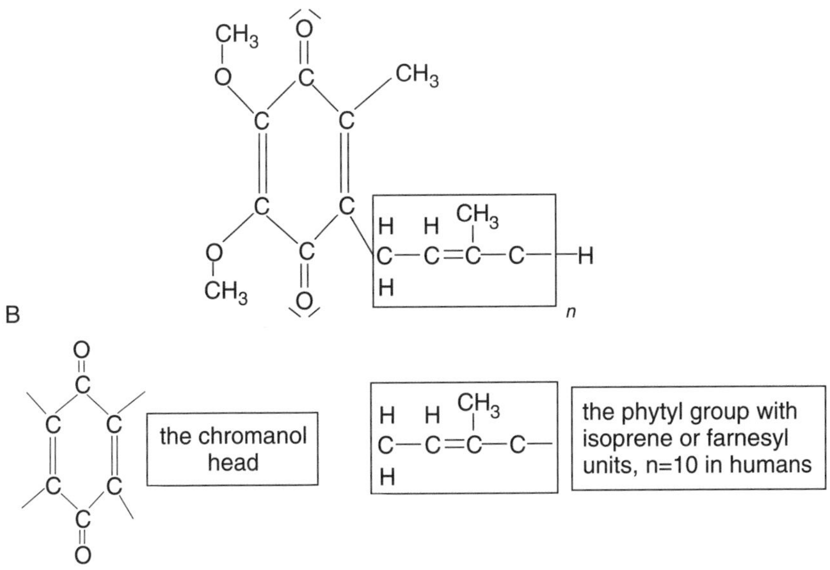

Figure 4.1b   The molecular structure of the endogenous quinone in human tissues, ubiquinone. The phenol- or benzene-ring structure (the phenol head) originates from the essential amino acid phenylalanine. The tail, in the form of a carbon chain, is comprised of isoprene (farnesyl) units produced by the endogenous mevalonate pathway. The phenol ring, isoprene chain, and corresponding entities in the vitamin E molecule are the chromanol head (ring) and the phytyl chain.

and Ingold 1993]. With respect to vitamin E, the double phenol-ring structure is referred to as the *chroman* or *chromanol head* and the carbon chain as the *phytyl unit* or *chain.*

Phenol Species and Their Scavenging Potentials

The aromatic structure implies unique biophysical properties to the molecule, which are modified in nature by, for example, methoxylation, glycosylation, and substitution. One such major feature for many phenol species is their lipophilic, which by definition also means their hydrophobic, property. Vitamins Q, E, A, and ß-carotene are soluble only in lipids and lipoidic structures.

In normal conditions, the major site for cellular radical formation is the lipid membranes of mitochondria, lysosomes, peroxisomes, and so on.

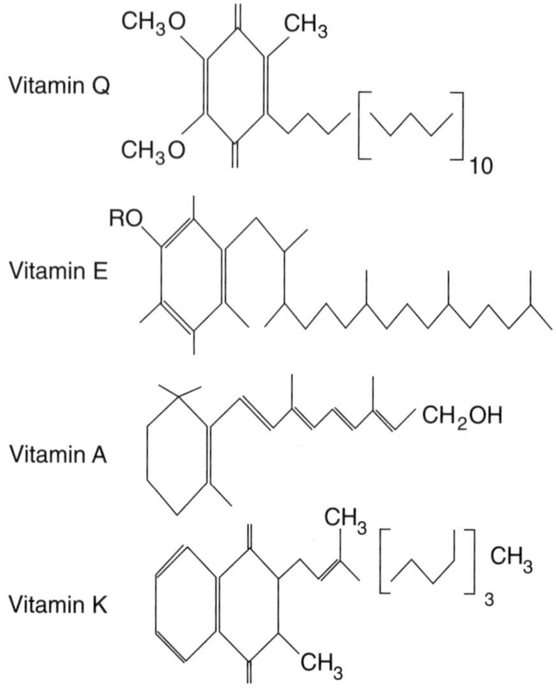

Figure 4.1c   The molecular structure of vitamins Q and E and two similar vitamins (A and K). Vitamin A is the precursor of 10 to 20 different signal substances related to cell growth. Vitamin K is necessary for blood clotting but may also be a precursor of the vitamin Q synthesis.

The first defense line against uncontrolled radical formation, based on these lipophilic antioxidants, is also located here.

Many recent investigations have found that the lipophilic antioxidant vitamins Q and E are interrelated. Vitamin Q exerts a certain role on vitamin E and can be referred to as a catalyst of the vitamin E reaction. In most tissues, vitamin E concentration exceeds that of vitamin Q by a factor of 5 to 20 [Stocker, Bowry, and Frei 1991; Karlsson, Diamant, Theorell, and Folkers 1993; Frei and Ames 1993; Esterbauer et al. 1993]. Reduced vitamin Q is continuously formed by mitochondria to regenerate vitamin E from its radical form—the alpha-(α-)tocopheroxyl radical [Maguire et al. 1992]. Vitamin Q is also more potent as a quencher, as it can act both at the initiation and propagation stages of lipid peroxidation; vitamin E, in contrast, quenches propagation reactions and subsequently breaks chain reactions [Ernster and Forsmark-Andrée 1993] (see figure 4.2). It is estimated that vitamin Q as an antioxidant is 10 times stronger than vitamin E [Stocker, Bowry, and Frei 1991].

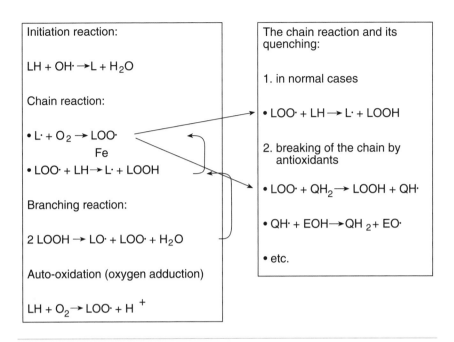

Figure 4.2   The main sequences of the cascade reaction and their quenching by antioxidant reactions.

In addition to these biochemical qualities, the biophysical qualities should be briefly explained, and they will be dealt with in greater detail in the next section. Vitamin Q is located within the lipid membrane and at the site of the oxyradical-producing enzymes, whereas vitamin E is allocated to the outer surface of the membrane. One could infer from this that the strategic position rather than the biochemical properties of vitamin Q explains the relationship between vitamin Q and vitamin E.

Radical Scavenging

Under normal conditions, most tissues in humans are predominantly in the reduced state. Reduced vitamin Q is continuously formed by the respiratory chain in the mitochondria (see figure 4.3). Reduced vitamin Q and reduced antioxidant compounds are both prerequisites for proper antioxidant function. Vitamin Q is in a stage of balance, or acquired equilibrium, with vitamin E and all the other phenol- and polyphenol-related compounds with antioxidant properties (according to the mass-action law, a formal equilibrium does not exist in the living cell). As a result, the mitochondria will feed the total antioxidant system—including

Mitochondrial metabolism and the reduced state
of different coupled antioxidant systems

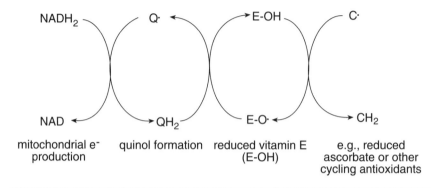

| mitochondrial e⁻ production | quinol formation | reduced vitamin E (E-OH) | e.g., reduced ascorbate or other cycling antioxidants |

Figure 4.3 The significance of electron generation by mitochondria and the coenzyme Q_{10} transfer system to maintain vitamin Q and other antioxidants in a reduced stage. This process is necessary for the antioxidant property of vitamin Q and of other antioxidants (e.g., vitamin E, vitamin C, and glutathione).

vitamin C, vitamin P, and glutathione—the necessary reducing equivalents through the function and position of vitamin Q as coenzyme Q.

Reduced quinones (quinols) in general react with free radicals, as described here:

1. The superoxide radical:

$$QH_2 + 2\ O_2^{\cdot -} \rightarrow Q + H_2O_2 + O_2$$

H_2O_2 (hydrogen peroxide) is metabolized in many different ways based on enzymatic process, and water plus molecular oxygen is regenerated. This is described in chapter 5.

2. An organic free radical from a lipid, protein, or nucleotide (as symbolized by R·):

$$QH_2 + 2\ R^\cdot \rightarrow Q + 2\ R$$

3. Peroxyl radicals:

$$QH_2 + 2\ ROO^\cdot \rightarrow Q + 2\ ROOH$$

The hydroxyl radical (OH·) is present only in the water phase. It is formed by the superoxide radical and water. Lipophilic quinones are not present in water; therefore, they do not scavenge this radical species or other exclusively water-soluble free radicals.

Water molecules are present in the lipid membranes but bound with different noncovalent bindings to lipoidic structures (*crystal water*).

The chain-breaking reaction with vitamin Q

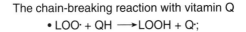

$\bullet$ LOO$\cdot$ + QH $\longrightarrow$ LOOH + Q$\cdot$;

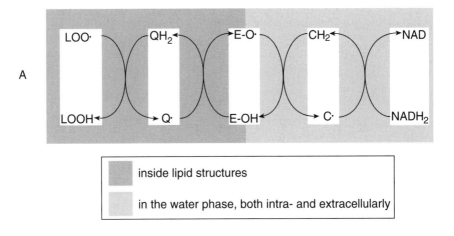

| | inside lipid structures |
| | in the water phase, both intra- and extracellularly |

The chain-breaking reaction with vitamin E

$\bullet$ LOO$\cdot$ + E-OH $\longrightarrow$ LOOH + E-O$\cdot$;

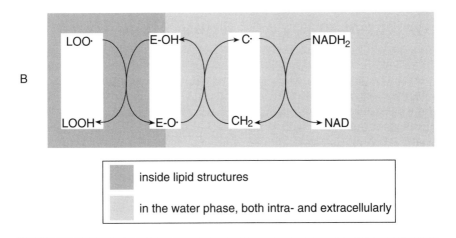

| | inside lipid structures |
| | in the water phase, both intra- and extracellularly |

Figure 4.4   (a) The lipid peroxidation chain reaction and its quenching, and the transport of the electron/radical to the cytosol and the water phase, where superoxide dismutase (SOD), catalase, and glutathione-dependent enzymes recover the electron for the respiratory chain and molecular oxygen (O_2) for water formation. (b) The same, but for vitamin E.

These molecules *could* be the target for the superoxide radical. The validity and extent of this theoretical possibility of hydroxyl radical formation has not been examined in its biological context.

Models as to how quinols react as antioxidants in different processes are also valid for the reduced vitamins Q, E, and C, and other hydrophilic antioxidant vitamins in the water phase.

When radicals and lipids are discussed, R is frequently substituted for L. Breaking according to 2. (L·) means scavenging at the initiation stage (see figure 4.4a), whereas 3. (LOO·) means scavenging at the propagation level in chain reactions (see figure 4.4b).

The antioxidant concept presented above is based upon the fact that the antioxidant is in a reduced state and donates electrons to a radical. Recently, it has been suggested that in underperfused tissues or in states in which there is lack of oxygen (i.e., hypoxic conditions) electron acceptance can take place instead of donation. That is, antioxidant activity can be obtained by the catalytic antioxidant vitamins acting as electron acceptors [Scott 1995]. The normal process by which L generates LOO in the chain reaction demands oxygen. But when oxygen is missing, L can react with the corresponding antioxidant vitamin radicals (see figure 4.5). Under these conditions semiquinone and the corresponding vitamin E radical (the phenoxyl radical) can be used experimentally to aid in the scavenging processes of LOO· . The biological implications of this biochemical property of vitamin Q and E radicals need further exploration.

Water Solubility and Antioxidant Allocation

As already mentioned, vitamins Q and E are lipophilic compounds; they are soluble only in lipids. The chromanol head of vitamin E, in contrast to the corresponding structure of vitamin Q, can electrostatically attract the dipolarized water molecule. The affinity of water molecules for the chromanol head means that this portion of the vitamin E molecule is water soluble.

The interaction between water molecules and the chromanol head tends to allocate vitamin E to the surface of the lipid membrane facing the water phase [Perly et al. 1985]. This is a competitive advantage of vitamin E over vitamin Q, which by definition suppresses vitamin Q in these particular compartments. In the inner part of the lipid membrane, however, they are equal in terms of lipid solubility. As many membranes constitute only lipid bilayers, it is obvious that more vitamin E will be dissolved than vitamin Q [Perly et al. 1985]. This is one possible explanation for the much higher content of vitamin E than vitamin Q in biological conditions.

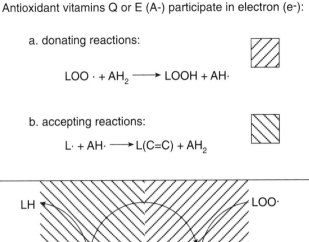

Antioxidant vitamins Q or E (A-) participate in electron (e-):

a. donating reactions:

$$LOO \cdot + AH_2 \longrightarrow LOOH + AH \cdot$$

b. accepting reactions:

$$L \cdot + AH \cdot \longrightarrow L(C=C) + AH_2$$

Figure 4.5   The antioxidant vitamins Q and E have been shown to be able to act catalytically, scavenging radicals in alternative reduction-oxidation (red-ox) reactions under conditions similar to oxygen lack (hypoxia). During hypoxia the antioxidant vitamins will be reduced by the actors in the initiation steps of the lipid chain reaction (figure 4.3 and 4.4a).

The allocation of the vitamin E molecule and its chromanol head also affects the antioxidant function both with regard to vitamin Q in the inner part of the lipid membrane and with regard to vitamin C and other hydrophilic antioxidants. The electrons obtained by the vitamin Q molecules in the scavenging process are transferred to vitamin E molecules. Vitamin E, in contrast with vitamin Q, can in turn transfer its radical originating electrons to vitamin C and to the other hydrophilic antioxidants. Thus, a certain fraction of the cell's vitamin C pool is compartmentalized to the lipid-water interfaces. The strength of this attachment at the molecular level is presently under investigation.

This fraction of vitamin C and its activity can be seen as a means to catalyze the scavenging activity of the hydrophilic antioxidant systems (the remaining vitamin C pool, vitamin P, glutathione, different antioxidant enzyme systems, etc). In this respect, vitamin C is similar to vitamin Q, which enhances the exploitation of the lipophilic antioxidant systems, primarily vitamin E and beta-carotene. This is the rationale for referring to this antioxidant strategy as the vitamin Q-E-C cycle. This is the basis for the nutratherapy programs suggested later in the book, in both preventive medicine, in general, and sports medicine, in particular.

Other Antioxidant Compounds

In addition to these cycling antioxidants, there are many other cellular compounds that have antioxidant properties [Draper 1993]. Some examples are glucose and oligosaccharides, urea, and bilirubin. Other molecules, such as polyunsaturated fatty acids and hyaluronic acid, react with and trap radicals.

Here are other nutrients with interesting antioxidant properties:

• Sulfur-containing organic compounds (allylic sulfides), such as sulfur amino acids in onion and garlic
• Indole alkaloids (indoles, "limonoids," e.g., benzpyrrole) in citrus fruits
• Monoterpenes in carrots

The significance of a mixed, well-balanced diet as the basis for a successful nutratherapy in sports medicine is obvious.

To say the least, antioxidant research has only scratched the surface in terms of the nutrients contained in food. Most probably, a number of antioxidants are yet to be discovered that are just as critical as the ones known.

Vitamin C also has a ring structure, although it is different from the benzene ring of the phenol species. But as is true for quinols, the molecule has hydroxyl groups adjacent to a double bonding between carbons. In vitamin C, too, the electron exchange takes place at this site of the molecule.

Glutathione also participates in red-ox reactions in the cell. But in this case the mechanism is based on another molecular feature—the tripeptide molecule, which contains a sulfur (S) section in the form of an L-cysteinyl residue. This structure is susceptible to dehydrogenation. Two glutathione molecules are then converted to a reduced disulfide form. The reduced form of glutathione is referred to as GSH and the oxidized form (glutathione disulfide) as GSSG:

$$2 \text{ GSH} \rightarrow \text{GSSG} + 2 \text{ e}^- + 2\text{H}^+$$

This reaction is regulated in part by the enzyme glutathione peroxidase (GPX) (see figure 4.6). GPX exists in two forms. One is dependent on the micromineral (trace element) selenium (Se) as a prosthetic group.

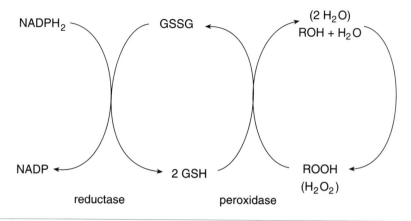

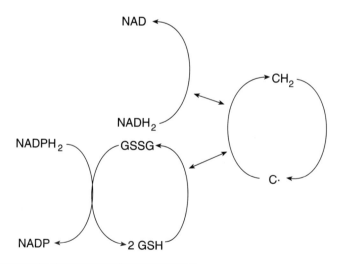

Figure 4.6   The oscillatory reaction of glutathione from its reduced stage (GSH) to its oxidized stage (GSSG) and back. The former reaction means an antioxidation activity. The enzyme glutathione reductase (GRD) catalyzes the reaction, where the reduced form (GSG) is formed by the electrons fed to the antioxidant system by the mitochondrial metabolism. The enzyme glutathione peroxidase (GPX) reduces the risks for hydrogen peroxide and other peroxides to translate into further uncontrolled reaction, whereby reactive radical species are formed.

Figure 4.7   Vitamin C is kept reduced and prepared as an antioxidant by glutathione and the glutathione-related enzymes as well as by vitamin Q. In both cases, electrons generated by the mitochondria are the means. The electrons are translocated by the NAD and NADP pools as cosubstrates.

The other form is independent of Se. The topics of GPX and antioxidant enzyme systems are discussed in more detail in chapter 5.

The red-ox reaction also depends on the enzyme glutathione reductase (see figure 4.6), which mediates the electrons from the electron-producing mitochondria. This will favor the presence of GSH over GSSG.

Both GSH and a more direct allocation of electrons from the mitochondria through reduced nicotinamide adenine dinucleotide (reduced NAD or NADH) and dehydrogenase enzymes maintain vitamin C in the reduced state (see figure 4.7).

Quenching of Singlet Oxygen

Although singlet oxygen ($^1O_2$) formally is not a free radical, its presence in the living cell is synonymous with a similar threat due to its energetic state (see chapter 2). Singlet oxygen has been shown to be capable of inducing DNA damage and to be mutagenic [Rånby and Rabek 1978; Sies et al. 1992].

Beta-carotene, in addition to its normal antioxidant activity, is designated to handle singlet oxygen. It is not a matter of supplying a reaction with electrons as a reductant (chemical scavenging), but rather to receive and "neutralize" the energy content that singlet oxygen represents in comparison with ordinary molecular oxygen (O_2) (physical scavenging). The size of the molecule—almost twice that of vitamins Q, E, and A (see figure 4.1c)—might explain the assignment:

1. $^1O_2$ + beta-carotene → O_2 + "energized" beta-carotene
2. energized beta-carotene → energy as heat

The beta-carotene molecule has the ability to receive all the energy and to release it in smaller "packages" (quanta) to its surroundings, which are not harmful to the cell and its constituents.

Other compounds involved in the scavenging of singlet oxygen, based on both physical and chemical scavenging processes, are vitamin E [Sies et al. 1992], amine, sulfide, and phenol species [Rånby and Rabek 1978]. These compounds, however, have a lower scavenging capacity than beta-carotene (50-fold less), a capacity that is even lower (100-fold less) than the beta-carotene-like bioflavonoid lycopene. Lycopene is abundant in tomatoes and gives the tomato its color.

Summary

Many cell constituents have antioxidant properties (i.e., they are reductants). The unpaired electron of the radical is scavenged by these red-ox reactions. Substituted phenols and polyphenols seem to be the major antioxidants.

These phenol-based molecules oscillate between two extremes—the reduced form, which is the antioxidant on alert (the quinol) and the oxidized and exploited form (the quinone).

A great number of variations on this theme exist in the living cell. The most frequent and central seems to be ubiquinone, coenzyme Q, or vitamin Q, which also is an endogenous synthesis product in humans. This compound is crucial and has catalytic properties with respect to the antioxidant activity of the exogenous antioxidant vitamin E. It is also essential to pump electrons into the antioxidant systems from the electron-producing mitochondria. The reduced state is a prerequisite for antioxidant activity. Vitamin Q has a central role to feed electrons into the antioxidant defense machinery.

The majority of free radicals are formed in lipid layers; therefore, the first antioxidant defense line is that of the lipophilic antioxidants vitamins Q and E. Vitamin E is present in much higher (5- to 20-fold) concentrations, which gives it a greater quantitative significance.

Vitamin E and its chromanol head has some affinity to water, in contrast to vitamin Q and its phenol head. This affinity allocates vitamin E toward the water phase of the lipid membrane and is a prerequisite for the mechanism to transfer radicals (electrons) to the hydrophilic and secondary antioxidant defense system, as represented by vitamin C. This transfer will drain the vitamin E pool of its radical-originating electrons and transfer them to the hydrophilic antioxidant systems, including vitamins C and P, the glutathione system with its enzymes GPX and GRD, and other antioxidant enzyme systems. The glutathione system will ultimately allocate the radical electrons to the respiratory chain. When applicable, even molecular oxygen will be fed into the respiratory chain to receive electrons and hydrogen to form water.

In the vitamin Q-E-C cycle, vitamin C can also be said to unload vitamin E on its radicals, thus providing the lipid layer with fresh vitamin E. Vitamin Q increases the efficiency of vitamin E's antioxidant activity. Vitamin Q is a catalyst of vitamin E and can also substitute for vitamin E in breaking lipid peroxidation chain reactions. Vitamin Q, but not E, can also scavenge initiation reactions of lipid peroxidation.

5

Antioxidants— Recycling Systems or Irreversible Reactions?

Free radicals exhibit a wide variety of chemical reactions. Some of them are very stable and do not represent any major threat to the living cells in the exercising muscles, whereas other radical species are extremely reactive and, therefore, harmful [Jenkins and Goldfarb 1993].

The major free radicals in lipid peroxidation can be ranked as follows (with respect to reactivity and thus aggressiveness):

1. The hydroperoxide radical $OH^{.}$
2. The carbon-centered radical $L^{.}$
3. The peroxyl radical $LOO^{.}$
4. The superoxide radical $O_2^{.}$

Many compounds in the living cell can react with a free radical (see table 5.1). The corresponding changes of the receiving compound are of vital importance for the cell (see figure 5.1). The compound might suffer structural changes that are incompatible with the expected function of the compound. The compound must retrieve its original structure or

Table 5.1   Solubility and Typical Concentrations (μM) of Compounds With Antioxidant Properties in Human Plasma (Stocker and Frei, 1991)

Compound	Concentration
Lipophilic antioxidants	
Vitamin E	25-40
Vitamin Q	0.4-1.0
Lycopene	0.5-1.0
β-carotene	0.3-0.6
Hydrophilic antioxidants	
Protein thiols	350-550
Uric acid	160-470
Vitamin C	30-150
Bilirubin	5-20

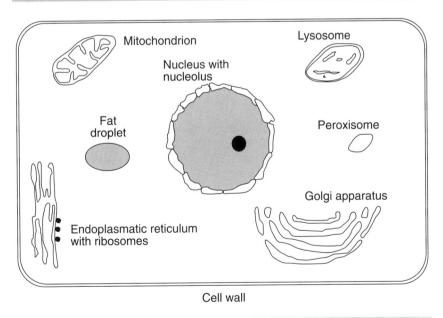

Figure 5.1   A schematic view of the cell, its organelles, and other cell parts, which constitute the lipid membranes, where the majority of enzyme processes take place and free radical species are generated and scavenged by antioxidants.

function or it could be expelled or excreted. Many of these compounds are present in plasma and can be analyzed there, which might have clinical relevance.

Design of Recycling Systems in Biology

In biology, chemical reactions, in which one or several products from one reaction participate in other reactions while regaining their original structure, have central roles in cellular metabolism. They are frequently referred to as *cycling reactions* because one or several participants oscillate between two extremes. Related to this oscillating behavior can be a chemical transportation of entities or parts from one molecule to another and from one compartment to another or through a membrane (see figure 5.2). The quinone-quinol reaction and the corresponding red-ox reaction in chapter 4 is one such example where electrons were shipped back and forth.

A schematic view of such coupled reactions in a cycling system might look like the following:

1. $A + B \rightarrow A' + B'$

2. $A' + C \rightarrow A + C'$

In this example, the compound A was altered (some parts were either lost or gained) in reaction 1. In reaction 2, A' was given back its original properties—A. The formed products B' and C' might be semi-products for further biochemical processes or they might be waste products. In the latter case, they might face an immediate or a later excretion from the cell. Other intermediate reactions might also be involved, and A is changed in different steps:

1. $A + B \rightarrow A' + B'$...

n. $A^n + C \rightarrow A + C'$

In reaction "n," the original compound A is regained.

These cycling reactions should not be mistaken for a reversible reaction, which in its simplest form can be described as in reaction 3:

3. $A + B \leftrightarrow A' + B'$

A' can be regained by running reaction 3 backward. If A' is permanently changed, the reaction is referred to as irreversible, and compound A irreversibly altered.

The term *reversible* is applicable to the compound A/A' in reactions 1 and 2. It can be suggested that compound A has been reversibly changed in the reactions and regained through *coupled reactions.*

In biology, both reversible and coupled reactions are exploited. Reactions 1 through n are the basis for transporting or shuttling reactions

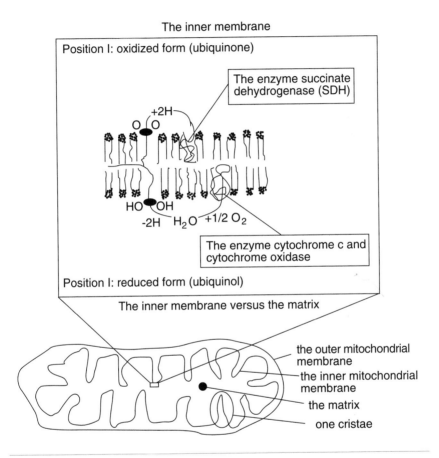

Figure 5.2   A single mitochondrion with its cristae. One section is cut out to depict the position of vitamin Q's two major parts—the chain and the ring—and the vitamin's function in electron transport between the three carbon acid (TCA) or citric acid cycle to the respiratory chain. The enzyme succinate dehydrogenase (SDH) of the TCA cycle donates two electrons to the oxidized form of vitamin Q (ubiquinone, depicted with 2 oxygens at each side of the ring). With this process, vitamin Q is reduced (ubiquinol formation, the 2 oxygens are changed to 2 -OH or hydroxyl groups) and the ring swings to the inside, where it meets the enzyme cytochrome oxidase of the respiratory chain. This enzyme takes over the two electrons, thus reoxidizing vitamin Q, whose ring now swings back to the other lipid layer to receive further electrons. The red-ox cycle is completed.

over membranes, for example (see figure 5.2). A' in reaction 1 penetrates the membrane with the transported entity and delivers it to the other side of the membrane. A' and other metabolites, but not A, has the ability to penetrate the membrane. Membranes usually contain a high proportion of lipids. Lipophilic properties of A', in contrast to A, can be the critical feature and allow diffusion through the membrane as a lipid-soluble (lipophilic) metabolite. Another feature is that A', but not A, can attach to a transporting vehicle such as a translocation protein and can pass through the membrane.

A further example of a shuttle reaction is transport of energy. ATP is the major cellular compound to retrieve and transport or allocate chemical energy from carbohydrate and fat sources to energy-depending processes in cellular metabolism. ATP is then used in different energy-depending processes. ATP is hydrolyzed to ADP and inorganic phosphate (P_i). Energy in ATP can also be stored ("buffered") by means of another high-energy phosphate compound, creatine phosphate (CP). From an exercise physiological point of view, ATP + CP are referred to as *phosphagens* [Karlsson 1971].

Vitamin Q as a Coenzyme in Mitochondria

Many biological reactions are based on red-ox reactions. Irrespective of the site of release of electrons, they will sooner or later end up in the mitochondria through different channels and will form water. The mitochondria consist of lipid membranes organized in layers. Vitamin Q will be found in the midst of the cristae lipids [Mitchell 1991]. The electron-handling property of the vitamin Q moiety, described earlier with respect to radical quenching, has been recruited as a traditional coenzyme to shuttle electrons between the TCA (three-carbon acid) or citrate cycle and the respiratory chain.

The Nobel Prize laureate Dr. Peter Mitchell has investigated and described the secrets of the mitochondrion, in general, and the electron transport, including vitamin Q as coenzyme Q, in particular [Mitchell 1991]. The mitochondrial membranes have five different layers: on each side of the outer membranes to prevent infiltration of water, inside the true lipid layers, and between them a layer of isolation, where the isoprene chain portion of vitamin Q is allocated. The remaining portion—the phenol ring—is found in the true lipid layers. The phenol head swings back and forth between the two lipid layers—from that facing the inner space of the mitochondria where vitamin Q is reduced (QH_2), to the outer layer, where water is formed and vitamin Q is oxidized by molecular oxygen (O_2).

It was estimated from studies in isolated mitochondria that approximately 30% of all cellular vitamin Q was allocated as a coenzyme in the mitochondria [Beyer, Nordenbrand, and Ernster 1986]. The first goal of coenzyme Q is to gather electrons from the different dehydrogenase enzymes in the mitochondrial cristae [Yu and Yu 1981]. The second goal is to establish a proton gradient across the membranes that can be coupled with ATP production [Mitchell 1991].

As discussed in chapter 4, the sequential red-ox processes of ubiquinone-ubiquinol include formation of the free radical semiquinone (Q). Semiquinone and water (H_2O) are the sources for superoxide formation [Boveris 1977]. Malfunction in the electron transport can lead to increased formation of the semiquinone radical. Chemicals that interfere with proper red-ox processes have been demonstrated to cause radical-generating dysfunction and deterioration in mitochondria [Crane, Sun, and Sun 1993].

Under normal conditions, about 3% to 5% of muscle oxygen turnover during exercise passes through the pool of free oxygen-centered radicals [Demopoulos et al. 1984]. These percentages may be underestimated for short, intense exercises where the electron pressure is elevated and lactate is formed [Karlsson 1971; Demopoulos et al. 1984]. In another study it was suggested that up to 15% of all oxygen might become a free oxygen radical [Sawyer 1988].

Antioxidants and Reversible Processes/Reactions

The relationships between the lipophilic antioxidant vitamins Q and E are examples of reversible reactions, where molecules are recovered through coupled reactions. The uniqueness of vitamin Q, with its access to the "proton-motive Q cycle" catalyzed by the ubiquinol-cytochrome c reductase of the respiratory chain, guarantees regeneration of ubiquinol both from the ubisemiquinone radical and ubiquinone (oxidized vitamin Q) [Mitchell 1976]. This will keep not only vitamin E reduced but also other parts of the antioxidant machinery, including hydrophilic systems, represented by vitamin C in the reduced state (see figure 5.3).

The reversible reactions and the ability to maintain vitamins Q and E in the reduced state are not only prerequisites to scavenge the cell of reactive radical species, but also prerequisites for not losing the antioxidant vitamins Q and E. If they stay oxidized they will ultimately be metabolized, and possibly decomposed [Packer and Fuchs 1993].

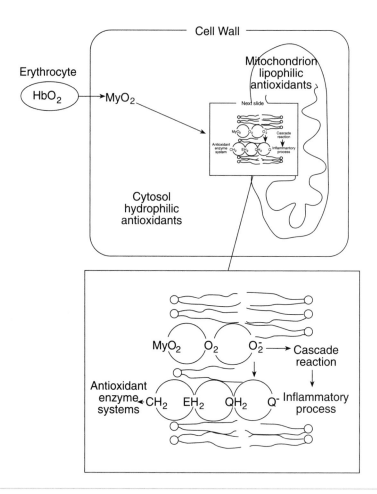

Figure 5.3   Peripheral oxygen delivery. Nighty-eight percent or more of molecular oxygen in arterial blood is bound to hemoglobin. In the capillary bed, the biochemical milieu favors a release. The concentration difference between arterial capillary blood and the surrounding tissue is large, which guarantees a fast diffusion for oxygen to the concentration-poor side. Intracellularly, myoglobin traps oxygen more efficiently than hemoglobin does because a higher affinity exists. Thus, molecular oxygen is free only during the very short passage from the capillary to the fiber. Otherwise, it is maximally protected by proteins to avoid uncontrolled radical formation. A similarly short period of "freedom" for molecular oxygen exists during the passage into the mitochondria and the respiratory chain—one of the main sources for an uncontrolled formation of free oxygen radical species. The majority of superoxide radicals formed, however, are quenched by ubiquinol (reduced vitamin Q). Vitamin Q can then directly or via vitamin E deliver the radical or the corresponding electron to hydrophilic antioxidants, such as vitamin C. In all these cases, with vitamins as antioxidants, the reduced forms $(QH_2, E\text{-}OH, \text{and } CH_2)$, are the active entity.

Irreversible Reactions

Radicals also participate in reactions which are irreversible. Some of these reactions are harmful to the cell—radical trauma is induced. Normal cell constituents and metabolites can be involved in these reactions. Can the moieties involved be referred to as antioxidants? The question might seem to be rhetorical, but under certain conditions it is relevant.

Polyunsaturated Fatty Acids

It is well documented that free radical species, uncontrolled by the antioxidant machinery, have polyunsaturated fatty acids as one of their targets. Their double bonds attract the unpaired electron of the radical species. Nerve tissue is rich in both saturated fatty acids and polyunsaturated fatty acids. Free-radical-generating experimental systems in a mixture of fatty acids peroxidize polyunsaturated fatty acids almost exclusively [Del Maestro et al. 1980] (see figure 5.4a). The peroxidized fatty acids cannot be repaired but instead are broken down to aldehydes and are excreted. Gases such as pentane and ethane are also formed and expired. Aldehydes can be determined in blood as thiobarbituric acid reacting substances (TBARS), such as malondialdehyde (MDA).

Polyunsaturated fatty acids are essential constituents of all membranes for many reasons. One is that their biophysical properties grant the membrane fluidity or plasticity [Drevon 1992; Dyerberg 1986; Leaf and Weber 1988], which is necessary for the contracting heart and skeletal muscle. Cell membranes of migrating white blood cells and the cell wall of red blood cells need polyunsaturated fatty acids to allow a smooth physical deformation in the passage through the capillary bed. PUFA-rich cell walls minimize or reduce hemolysis, whereas PUFA-poor cell walls, with their rigid, "stiff" structure, are more susceptible to mechanically induced rupture and hemolysis [Beving, Petrén, and Vesterberg 1990; Beving, Tedner, and Eriksson 1991].

White blood cells are the most antioxidant vitamin-rich cells known to science [Bendich, 1993]. Antioxidant vitamin-poor white blood cells and subsequent peroxidation of vitamin F_1 is most probably the main reason for the impaired immune function in elite endurance athletes [Green et al. 1981]. Runners anemia, well known to many elite endurance athletes, is also related to the increased vitamin F_1 susceptibility in athletes who are lacking in antioxidant vitamins [Packer 1986].

Polyunsaturated fatty acids are also a prerequisite for receptors to function in the membrane as expected. This is true for electrolyte transport in the heart muscle membranes [Hallaq, Smith, and Leaf 1992], for surface receptors of the white blood cell [Berg Schmidt et al. 1991], and for oxygen release mechanisms of the red blood cell [Cunnane et al. 1989; Tengerdy 1989].

Polyunsaturated fatty acids are usually bound to phospholipids in membranes. A portion of PUFA—the essential fatty acids, or vitamin F—is also a precursor for formation of the hormone group *eicosanoids*. They exist in different series: prostanoids (prostaglandins), leukotrines, and thromboxanes originating in vitamins F_1 and F_2. They appear in relation to and within the sites of inflammatory processes. A prerequisite for eicosanoid synthesis is release of vitamin F from the phospholipids. A PUFA release from these depots is achieved and regulated by the enzymes phospholipase A_2 and C [Pruzanski et al. 1985]. It is not possible to determine the presence of the enzyme phospholipase A_2 in clinically healthy subjects in contrast to patients with inflammatory conditions such as rheumatoid arthritis [Pruzanski et al. 1985].

The phospholipase enzymes will release large amounts of PUFA. Some of them will react with and trap free radical species as a result of their elevated plasma and extracellular levels. It is conceivable that the radical trapping by PUFA could protect more vital cell parts and tissues from these reactive radical species. Vitamin F_1 (omega-3 fatty acids) has a higher affinity to reactive radical species than vitamin F_2 (omega-6 fatty acids). It can also be expressed, as in previous sections of this book, as a higher susceptibility to radical trauma for vitamin F_1 than for vitamin F_2.

Hyaluronic Acid

In the synovial fluid of the joints, in heart and skeletal muscle, hyaluronic acid (HYA) is present as an essential part of the extracellular matrix. HYA has a significant effect as a lubricant and adds to the elastic properties of the muscle. In muscle, HYA provides a basis for energy storage in elastic components, which can be retrieved in the next muscle contraction. This form of mechanical energy may be added to mechanical energy derived from ATP splitting and transformation of chemical energy into mechanical work. This combined mechanical energy is important for the left ventricle of the heart, for the jumper's "take off" and for the long-distance runner's mechanical efficiency [Piehl-Aulin et al. 1991; Engström-Laurent and Haagren 1987; Komi 1987].

Hyaluronic acid reacts with free radical species and gets depolymerized and defragmented [Ng et al. 1992; Artola et al. 1993] (see figure 5.4b). Under conditions similar to inflammatory processes, the release of HYA is exaggerated. The corresponding HYA reactions with radicals will most probably mask and protect other vital cell parts and tissues [Engström-Laurent and Häägren 1987].

DNA, Cell Growth, "Supercompensation," and Radicals

Adaptation to increased physical demands—training effects—are essential both in sedentary people and in elite athletes. The nature of these processes, frequently referred to as "supercompensation," is not known and is probably multifactorial.

It is possible to conclude, however, that stimulation of gene expression in DNA, both in the cell nuclei and in the mitochondria, is critical to achieve the *de novo* synthesis of proteins. The cytoskeletal system in the muscle, for example, including acto-myosin and other filament proteins, is one example, and increased respiratory potential and mitochondrial enzyme activity is another. The former is governed by nuclear DNA (n-DNA) and the latter by mitochondrial DNA (m-DNA) [Luft 1994].

Both DNA pools are susceptible to radical injury (see figure 5.4c). It is well documented that lipid peroxidation may cause DNA damage referred to as *strand breaks* [Halliwell and Gutteridge 1986; Alessio et al. 1990; Alessio 1993]. Recently, it was documented that DNA is affected by physical exercise, which simultaneously led to radical formation of the kind and order known to initiate lipid peroxidation [Inoe et al. 1993].

In fact, lack of vitamin Q has been indirectly confined to DNA-related disorders such as Luft disease [Luft et al. 1962], myopathies and neuropathies [Ernster and Lee 1990], and other forms of oxidative tissue injuries [Ames 1989].

Age-related changes in the endogenous synthesis and allocation of vitamin Q are most likely related to changes in DNA. In a series of studies, Gustav Dallner and coworkers revealed these changes for different tissues in humans [Kalén, Appelkvist, and Dallner 1989; Dallner 1994; Appelkvist, Kalén, and Dallner 1991]. Peak values in the heart and liver seem to occur at the age of 20 years and then decline (see figure 5.5). This contrasts with the contents of the gland tissue vitamin Q (the pancreas and the adrenals), where the contents peak at a very young age (about 2 years) and then decrease.

Vitamin Q, Mitogenic Ligands, and Training

Heavy physical exercise is known to elevate both muscle and blood lactate and pyruvate concentrations [Karlsson 1971; Karlsson, Rosell, and Saltin 1972]. But it is equally well recognized that the blood lactate increase precedes the blood pyruvate increase [Karlsson and Saltin 1971]. Moreover, evidence is present that lactate might be partly oxidized by

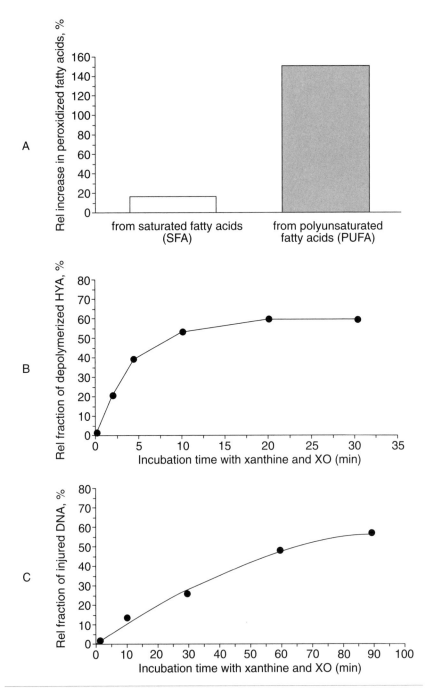

Figure 5.4 (a-c) Experimental radical formation and the effects on saturated and polyunsaturated fatty acids in brain ischemia (experimental stroke), hyaluronic acid (HYA), and DNA [Del Maestro et al. 1980].

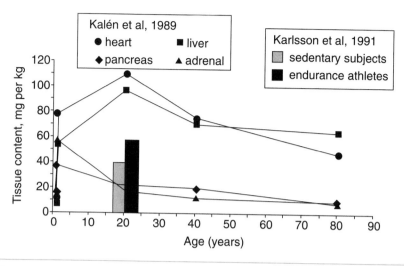

Figure 5.5. Vitamin Q in heart muscle, liver, and secretory glands (pancreas and adrenal) with age in humans. For comparison, skeletal muscle *(m vastus lateralis)* vitamin Q for young sedentary males and male endurance athletes is included.

muscle tissue at the membrane level resulting in a venous efflux of pyruvate [Karlsson, Rosell, and Saltin 1972]. This would indicate membrane-bound red-ox processes, which lead to an inward electron transport [Jöbsis 1964].

It is now established that the cell membrane contains a vitamin Q-coupled NADH reductase to transport electrons over the membrane. Moreover, membrane-bound vitamin Q participates in the control of internal pH and calcium exchange [Crane et al. 1991b; Crane et al. 1993]. This membrane-bound electron transport stimulates the cell growth in the same way other cell growth stimulating (mitogenic) factors do. Mitogenic factors (ligands) include, in addition to pH and calcium (Ca^{2+}) changes, features such as activation of protein kinases and proto-oncogenes [Crane et al. 1991b; Crane, Sun, and Sun 1993]. Cytometric analysis has shown that vitamin Q enhances DNA synthesis and transition [Crane, Sun, and Sun 1993].

It is obvious that the occurrence of these reactions is a prerequisite for an intact muscle membrane quality. Trauma to the membrane by free radicals can occur when antioxidant levels are depressed or cannot meet the needs to protect from an exaggerated radical formation. Polyunsaturated fatty acid peroxidation will then result and can lead to reduced effectiveness of the cell membrane in, for example, skeletal muscle. This process can result in reduced muscle function in fitness athletes as well as in elite athletes. Supercompensation processes, as a result of fitness or elite athletic training, may also be reduced or even eliminated as a result of radical trauma.

Cycling Antioxidants and Antioxidant Enzymes

It is evident that the lipophilic antioxidant vitamins Q and E participate in reactions, where the radicals or rather the unpaired electron will be transferred to enzymes through cycling reactions. These enzymes will catalyze reactions, whereby molecular oxygen (O_2) and electrons (e^-) will be forced to enter the more normal and safe reactions leading to formation of water (H_2O).

Superoxide Dismutase-Catalase System

In most cases, the mitochondrial "spillover" of the superoxide radical represents a normal and relatively safe formation of radicals. Radical intermediates are necessary semi-manufactures or by-products in many synthesis processes. This spill of the superoxide radical is most probably a biologically "planned" source of radicals. Superoxide formation is surrounded and guarded by the antioxidants in a manner similar to that of safety measures for nuclear plants: safety first. In fact, some biochemists claim that the superoxide radical is a "safe" radical [Sawyer 1988].

In addition to vitamins Q and E, another major antioxidant system is the enzyme system with antioxidant properties. The antioxidant enzyme system is based on superoxide dismutase (SOD), catalase (Cat), and glutathione peroxidase (GPX). These primary enzymes have support enzymes such as glutathione reductase (GRD), glucose-6 phosphate dehydrogenase (G-6-PDH), and glutathione sulfur (S) transferase (GST), which supply the enzymes with reducing equivalents and substrates [Ji 1993].

The enzyme superoxide dismutase catalyzes the reaction:

$$2\ O_2^{\cdot} + 2\ H^+ \xrightarrow{\text{SOD}} H_2O_2$$

The SOD enzyme exists in different protein forms, which have the following metallic prosthetic groups [Deby & Goutier 1990]:

- A protein with manganese (Mn) at the protein-active site and as a prosthetic group (see table 2.4) (Mn-SOD, SOD_{Mn}), located in the mitochondria matrix and close to the site where the superoxide radical is formed—the succinate dehydrogenase-cytochrome b segment. This enzyme is coded by the mitochondrial DNA pool (m-DNA) [Nohl 1986].
- A protein with copper (Cu) and zinc (Zn) at the active site (Cu-Zn-SOD, SOD_{CuZn}), located in the cytoplasm and in the extracellular fluids

As mentioned previously (see figure 2.2a), muscle SOD activity in humans increases with either maximal pulmonary oxygen uptake ($\dot{V}O_2$max) or percent distribution of slow-twitch muscle fibers (%ST) in the thigh muscle (*m vastus lateralis*) [Jenkins, Friedland, and Howald 1984]. Hydroperoxide (H_2O_2), although not a radical, can cause autooxidation, which will be dealt with in chapter 6. Hydroperoxide is trapped and "neutralized" by another constituent of this antioxidant enzyme system—catalase (Cat), as seen in the following:

$$2\ H_2O_2 \xrightarrow{\text{CAT}} 2\ H_2O + O_2$$

In general, peroxide moieties can also be "repaired" by this enzyme in the following reaction:

$$ROOH + AH_2 \xrightarrow{\text{CAT}} RDH + H$$

The catalase enzyme is also a mitochondrial enzyme but is present primarily in other organelles, such as peroxisomes.

Glutathione Peroxidase and Reductase Systems

Glutathione is a predominant nonprotein sulfhydryl (SH)-containing compound in living organisms. The SH groups are provided by the essential amino acid cystein.

In chapter 4, the enzyme glutathione peroxidase (GPX) was said to catalyze red-ox reactions. The enzyme GPX exists in two protein forms as well. One is dependent on selenium (Se-GPX, GPX_{Se}) as a prosthetic group, and the other is nondependent on selenium. GPX catalyzes the reaction, where organic and inorganic hydroperoxides are the substrate and water is formed.

$$2\ ROOH \xrightarrow{\text{GPX}} 2\ ROH + H_2O;$$

This is coupled with the reversible reaction:

$$GSSH + 2\ e^-; \xleftrightarrow{\text{GRD}} 2\ GSH$$

This red-ox reaction is catalyzed by the enzyme GRD (see figure 4.6). The electrons originate from electron-producing red-ox processes in the mitochondria with vitamin Q as a critical constituent in the shuttling mechanisms.

GRD can also help to maintain vitamin C in the reduced state (see figure 4.7).

When the pro-oxidant cellular milieu is enhanced, the tissue starts to release GSSG to the blood. The ratio GSH over GSSG (GSH x GSSG^{-1}, GSH/GSSG) is a clinical tool to study oxidative stress in fitness exercise participants, elite athletes, and patients.

Summary

Many compounds in the living cell are designated to react with free radical species as antioxidants. These antioxidants are present in all cells and tissues and are on the alert on a continuing basis.

Antioxidants such as vitamins Q, E, and C participate in coupled reactions, which will deliver fully recovered and reduced vitamins Q, E, and C as end products. They can then reenter reactions to scavenge newly formed radicals—the vitamin Q-E-C cycle. This is an example of recycling processes, which are frequent in many biological systems.

The coupled reactions with stepwise modifications of molecules and different affinity patterns to different transport vehicles are also the basis for shuttle mechanisms. Such shuttle mechanisms are vital for the vitamin Q fraction acting as coenzyme Q in the mitochondria to form ATP and also to regenerate reduced vitamin Q (ubiquinol), which will maintain all the other coupled antioxidant vitamin systems reduced.

Free radicals also react with more vital parts of the cell, and this leads to irreversible and perhaps permanent damage. Polyunsaturated fatty acids, hyaluronic acid, and gene expression in the form of DNA and RNA, in both the nuclei and mitochondria, are the targets for these reactive radical species.

Two enzyme systems are involved in the scavenging process: superoxide dismutase (SOD)-catalase (CAT) and glutathione peroxidase (GPX)-reductase (GRD). They are linked to the radical formation and its scavenging in a later and final stage of the total antioxidant process.

The antioxidant strategy is based on

1. the first defense line: the lipophilic antioxidants,
2. the cytosol and the water phase with its hydrophilic antioxidants, and
3. the antioxidant enzyme systems.

Through these processes, electrons are passed to the respiratory chain, molecular oxygen is recovered, and peroxide molecules are repaired.

6

Radical Formation in Different Cells and Tissues

Oxygen-centered radical formation in biological conditions can be summarized as follows:

1. *Electron transport chain and oxidoreductases.* Mitochondrial and microsomal functions, membrane-bound NADH and NADPH oxidases, and various red-ox enzymes produce the superoxide radical (O_2^-) and hydrogen peroxide (H_2O_2).

Heart ischemia, ATP degradation, and xanthine dehydrogenase (XDH) conversion to xanthine oxidase (XO) are the bases for superoxide and hydroxyl radical formation in infarction and subsequent cell trauma. There is controversy over whether these events can also take place in skeletal muscle.

2. *Auto-oxidation.* (a) Transition metals catalyze auto-oxidation of intermediary metabolites with appropriate red-ox potential such as sulfhydryl (SH-, thiol) compounds, enediols, and diphenols. (b) Adduction of free molecular oxygen to polyunsaturated fatty acids, for example, also must be considered.

3. *Decompartmentalization.* After partial destruction of compartment membranes, "unspecific" oxidations may occur due to the combination of substrates and enzymes normally separated in the intact cell. These reactions represent a stage of deterioration, necrosis, and decomposition of the radical-injured cell and the cell's death that could lead to infiltration of macrophages and neutrophils and subsequent radical formation.

4. *Inducible enzymes and xenobiotic or drug coupling.* The enzyme system P_{450} is present as a gene transcript in most cells and tissues. Providing ample substrates, the gene and corresponding protein synthesis is induced. These enzymes frequently lack specificity at their reducing sites and can be a cause of free radical formation.

In muscle contraction, radical formation occurs according to point 1 above, and can be related to the following physiological mechanisms:

- Radical formation related to an electron "leakage" or "spillover," physically located at the coenzyme Q_{10}-cytochrome b system in the mitochondrion
- An elevated electron pressure, perhaps in combination with accumulation of metabolites such as adenosine and conversion of XDH to XO. This provides the necessary mechanisms to allow adenosine to be oxidized to xanthine and then to uric acid or urate. In these last reactions, the superoxide or hydroxyl radical is formed by XO.
- Adduction of free molecular oxygen to double bonds such as in PUFA and a subsequent auto-oxidation (cf. "oxygen burst injury," the early state of circulatory shock, pure oxygen ventilation such as divers have, artificial ventilation including incubator treatment of premature babies, and ventilation support for patients with chronic obstructive pulmonary disease)

Muscle Activity and Mitochondrial Metabolism

It is estimated that up to 15% of the molecular oxygen turnover in cellular metabolism can pass through an oxygen-centered radical species and appear as a leakage of singlet oxygen or hydrogen peroxide [Sawyer 1988].

In exercising muscle, the radical formation estimate is in the order of 3% to 15% [Demopoulos et al. 1984]. Muscle radical formation can occur as a result of electron leakage or ischemia-reperfusion (see figure 6.1), which in molecular cardiology, for example, is referred to as the *oxygen or respiratory burst injury* or *reperfusion injury*. This radical formation condition can occur as the result of artery-clamping-induced ischemia, fermentative processes, capillary dilation, clamping release, and a subsequent immediate tissue oxygenation.

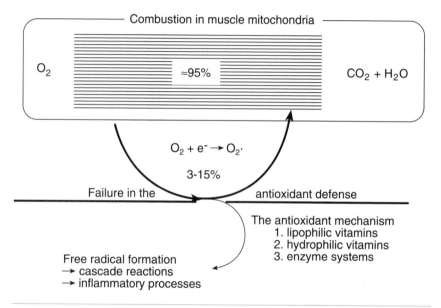

Figure 6.1   The electron leakage and subsequent formation of oxygen-centered free radicals, predominantly the superoxide radical (O_2^-), as a result of the electron transition from the citrate or tricarboxylic acid (TCA) cycle to the respiratory chain and the site of coenzyme Q_{10}-cytochrome B. Three to fifteen percent of the total turnover of molecular oxygen is assumed to pass through the pool of oxygen-centered radicals.

In general, oxygen-centered radicals can be formed by two means:

1. Leakage in metabolic pathways of electrons and energetically exaggerated metabolites. This phenomenon has been of major interest to biochemists and physiologists [Ernster and Beyer 1991].
2. Physically dissolved molecular oxygen in (a) transition from the capillary bed to the cell, and (b) the passage into the mitochondrion. Whether or not oxygen adduction is a possible source of radical formation in the living organism is being debated. Theoretical possibilities for such a mechanism, however, have been introduced in biology [Widmark 1993; Widmark 1957].

Regulation of Local Oxygen Transport

Under normal conditions, oxygen availability or tension in muscle is kept at an extremely low level. At the mitochondrial level, pO_2 equals a fraction of 1 mmHg. This is true for both heart and skeletal muscle. The healthy heart has an oxygen store, which at the most can cover four to six active beats. Resting and contracting skeletal muscle is always on the borderline to lactate release, which indicates a potential oxygen shortage [Saltin et al. 1968; Saltin 1985]. This metabolic milieu also grants a reduced state for antioxidants [Karlsson 1971; Karlsson et al. 1972; Karlsson, Diamant, Theorell, Johansen, and Folkers 1993; Karlsson 1987].

The increase in oxygen turnover with heavy exercise is at the muscle level of 10,000% to 20,000%. At the pulmonary level, the corresponding level is 1,000% to 2,000%, depending on the level of physical conditioning (see figure 6.2) [Saltin 1985; Savard, Kiens, and Saltin 1987]. Capillary dilation and increased local and central blood flow are strictly regulated with the local metabolic milieu in the contracting muscle as one driving force [Karlsson 1986b; Shepherd et al. 1981].

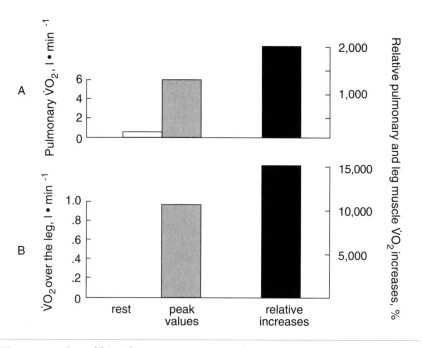

Figure 6.2 (a and b) Pulmonary oxygen uptake and the calculated oxygen uptake during one-legged exercise based on local blood flow and the oxygen arterial-venous difference [Saltin, 1985; Savard, Kiens, and Saltin 1987].

At the capillary level, dilation seems to be an all or nothing response (intermittent ischemia); when fermentation metabolites have accumulated to a certain critical level, the precapillary sphincters are opened [Mellander 1981]. The uncharged radical species nitric oxide (NO·) and adenosine are two other local metabolites that may be key factors [Shimokawa and Vanhoutte 1988; Bassenge 1992; Vanhoutte 1991].

It is suggested that a short nervous loop may exist between the capillaries of the contracting muscles via the spine or at a level beneath, resulting in smooth muscle relaxation of the arterioles for that particular muscle or muscle section [Mitchell and Schmidt 1983; Mitchell et al. 1977]. In addition to these local features, ergoreceptors sensitive to local muscle metabolism can activate central circulation through Type III and IV afferents (see figure 6.3, a-c) [Shepherd et al. 1981]. In humans, the ergoreceptor concept has been documented experimentally by means of registration of the local muscle metabolism (nuclear magnetic resonance spectroscopy, NMRS), efferent and afferent nerve traffic, and circulation changes [Henriksen et al. 1988; Victor et al. 1988].

Oxygen extraction of arterial, nutritional blood is 80% or more in muscle tissue. This figure will be compared with extraction of fuels as

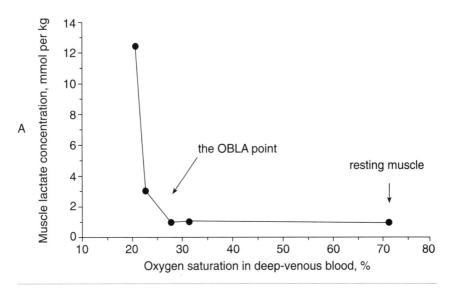

Figure 6.3a   The relationship between deep-venous oxygen saturation (SO_2) and muscle lactate concentration with cycle ergometer exercise at different intensities [Karlsson 1979].

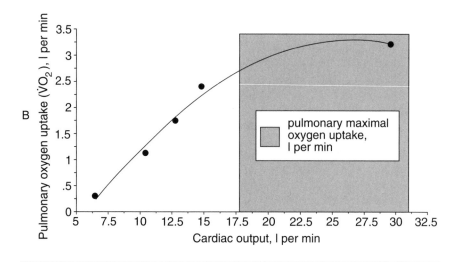

Figure 6.3b   The relationship between pulmonary oxygen uptake ($\dot{V}O_2$) and cardiac output during cycle ergometer exercise at different intensities [Saltin 1985].

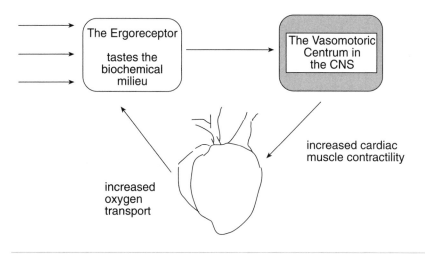

Figure 6.3c   A schematic description of regulation of central and peripheral circulation based upon the ergoreceptor concept, as introduced by Shepherd and coworkers [Shepherd et al. 1981]. The ergoreceptors are sensitive to the actual metabolic environment and fire directly to the vasomotor center in the central nervous system.

free fatty acids (FFA), glucose, and lactate, which rarely are higher than 20%. This almost maximal O_2 extraction illustrates the level of caution with which molecular oxygen is treated.

Taken together, in resting and contracting muscle, oxygen availability is tightly monitored and regulated. The teleological implication of these regulatory measures is to reduce the risk for oxygen radical species formation to an absolute minimum.

Skeletal Muscle Fiber Concept

As late as in the 1980s, to many exercise physiologists, muscle contraction was not more than a "sink" for molecular oxygen. In the early and mid-1970s the first scientific contributions appeared, which linked maximal pulmonary oxygen uptake ($\dot{V}O_2$max) to the quality of the muscle expressed as the oxidative capacity (see figure 6.4a) [Berg et al. 1978; Forsberg et al. 1976]. At first, these relationships were only appreciated as "coincidental." No further biological or physiological consequences of these relationships were recognized.

In the middle and late 1960s, animal exercise physiologists claimed that functional relationships existed between the periphery, expressed as muscle tissue biochemical properties, and central circulation, expressed as oxygen uptake [Holloszy 1967]. At that time, these data were neglected and explained as "species variations."

By the late 1960s, it was accepted that human skeletal muscle, in terms of fibers, was a mixed tissue with two or more major muscle fiber types that could be histochemically, physiologically, and neurologically identified (see figure 6.5a). Gollnick suggested the following nomenclature in their breakthrough studies [Gollnick et al. 1972]:

1. "Red" Type I, or slow-twitch (ST) muscle fiber
2. "White" Type II, or fast-twitch (FT) muscle fiber

The following features comprise the major differences between the two fiber types:

- The ST fiber is adapted to respiration and oxygen utilization, whereas the FT fiber is not [Armstrong et al. 1972].
- The ST fiber contains more myoglobin than the FT fiber does [Moll and Bartels 1968; Biörck 1949].
- The ST fiber has a more potent antioxidant machinery than the FT fiber has [Jenkins et al. 1984; Karlsson 1987].
- The ST fiber is surrounded by more capillaries than the FT fiber is [Andersen 1975].
- The ST fiber has a different ergoreceptor control promoting central and peripheral oxygen delivery, unlike the FT fiber [Shepherd et al. 1981; Mitchell 1985; Karlsson 1986a].
- ST fiber metabolic profile favors lactate respiration, unlike the FT fiber.

The ST fiber, however, unlike the FT fiber, lacks speed and strength [Sjödin 1976]. The FT fiber is provided with a different biochemical regulation of the metabolic pathways. Among other things, this results in a faster ATP hydrolysis and recharging of the ATP battery, whereas the ST fiber and its respiratory profile regenerate ATP more efficiently per unit of chemically bound energy transferred (see figure 6.5b) [Karlsson 1971; Komi and Karlsson 1979; Karlsson 1979].

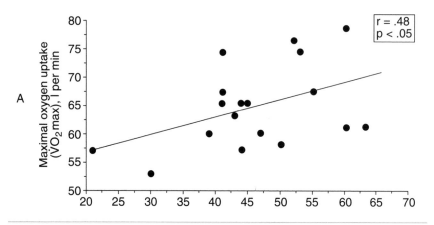

Figure 6.4a   The individual relationship between maximal pulmonary oxygen uptake and muscle fiber composition expressed as percent distribution of slow-twitch muscle fibers (%ST) in physically young males [Berg et al. 1978].

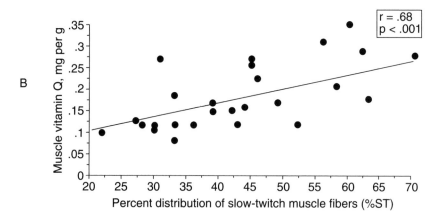

Figure 6.4b   The individual relationship between thigh muscle (*m vastus lateralis*) vitamin Q content and muscle fiber composition expressed as percent distribution of slow-twitch muscle fibers (%ST) in physically fit young males [Karlsson 1987].

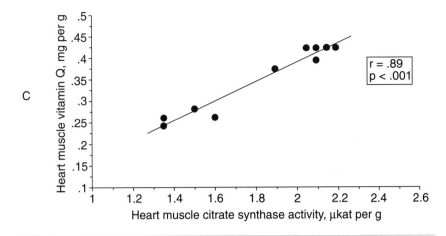

Figure 6.4c   The relationship between heart muscle vitamin Q content and heart muscle citrate synthase (CS) activity in healthy persons [Lin et al. 1988]. Each point represents the mean of six samples.

It is well documented that athletes participating in endurance sports, such as long-distance running, swimming, and cross-country skiing, have a high proportion of ST fibers (60% to 100%) in their contracting muscles as compared to those participating in short, exhaustive events or strength-dependent activities [Karlsson, Frith, et al. 1974; Karlsson et al. 1975; Karlsson 1979].

As with heart muscle, the oxidative capacity of skeletal muscle in humans covaries with the vitamin Q content (figure 6.4b and c) [Lin et al. 1988; Karlsson 1987].

Muscle Mitochondria, Exercise, and Radical Formation

Skeletal muscle mitochondria is perhaps the major site for radical formation, in general, and oxygen-centered radicals, in particular, during physical exercise [Witt et al. 1992]. In chapter 5 it was demonstrated how formation of the semiubiquinone radical and contact with water led to formation of the superoxide radical ($O_2^{\cdot}$) (see figure 5.4). There is no reason to speculate that muscle fiber composition or endurance training adaptation would have any relevance to this electron leakage *per se* [Witt et al. 1992; Jenkins, Krause, and Schofield 1993].

It has been demonstrated that endurance-type exercise in rats uses both vitamins E and C [Packer 1986]. In animal experiments, a great number of deficiency studies have been carried out demonstrating radical formation and skeletal muscle tissue damage following exercise [Witt et al. 1992]. The relevance of these studies to healthy humans, however, is not always obvious.

Trained rats have been reported to have less skeletal muscle mitochondrion lesions than untrained rats following a selenium-poor diet [Ji 1993]. Comparative studies have revealed that rats with sufficient vitamin E are more protected than rats with sufficient selenium. The antioxidant system involving vitamin E in these animal experiments is perhaps more important than the selenium-glutathione peroxidase system for radical-induced lesions. Moreover, endurance performance is relatively more increased than peak performance and peak pulmonary oxygen uptake in the animals after supplements were given [Gohil et al. 1991; Witt 1992].

The question has arisen as to whether antioxidant supplements given to animals reduce the exercise-induced muscle trauma that is most likely related to reactive radical species. In a rat study, vitamin E was given in pellets (*per os*). Soleus muscle vitamin E content increased 350% as compared to control animals. Reduced radical

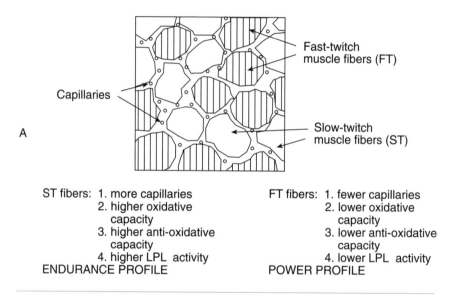

ST fibers: 1. more capillaries
          2. higher oxidative
             capacity
          3. higher anti-oxidative
             capacity
          4. higher LPL activity
ENDURANCE PROFILE

FT fibers: 1. fewer capillaries
          2. lower oxidative
             capacity
          3. lower anti-oxidative
             capacity
          4. lower LPL activity
POWER PROFILE

Figure 6.5a   Skeletal muscle in humans contains two major muscle fibers in the matured stage: Type I, slow-twitch (ST or "red") and Type II, fast-twitch (FT or "white") muscle fibers [Karlsson 1979]. These main muscle fibers have different biochemical and metabolic properties and different intracellular excitability processes, are innervated differently, and have different capillary network supplies including intracellular oxygen transport capacities. These differences mean that the ST fiber is the one designed for endurance, and that FT fiber is designed for short, explosive muscle activity.

1.   Mitochondrial ATP
      resynthesis

2.   2 ADP → ATP + AMP

      AMP-DA
3.   AMP → IMP

      5'-Nase
4.   IMP → inosine

B

      PN-phlase
5.   inosine → hypoxanthine

6.   hypoxanthine (HX) $\xrightarrow{xo}$ xanthine (X) + H_2O_2

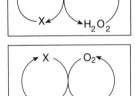

7.   xanthine (X) $\xrightarrow{xo}$ uric acid (UA) + $O_2\cdot$

Figure 6.5b Muscle contraction means that the chemical energy in ATP stored in the muscle as phosphagens (ATP + CP) is converted into mechanical work [Karlsson 1971; Komi and Karlsson 1979; Karlsson 1979]. ADP is recharged to ATP with the local CP stores and from mitochondrial ATP resynthesis. Under situations similar to severe ischemia in the heart muscle (e.g., myocardial infarction), AMP is accumulated and deaminated (AMP-DA, AMP-deaminase). Provided that xanthine dehydrogenase (XDH) is converted to xanthine oxidase (XO), hydrogen peroxide (H_2O_2) and the superoxide radical ($O_2\cdot$) are formed from hypoxanthine and xanthine, respectively. 5'-Nase = 5'-nucleotidase, PN-Phlase = purine-nucleoside phosphorylase.

formation was demonstrated by means of chemoluminiscence but had no impact on muscle histochemistry. Vitamin E supplements had no "saving effect" on rat muscle physical performance traits as studied *in situ* [Warren et al. 1992]. No muscle data on vitamin Q were provided in that paper. With current knowledge concerning the interplay between vitamins Q and E, it would be interesting to see the effect of a combined supplement program.

In a series of experiments, Tiidus and colleagues failed to demonstrate any impact on exercise performance in rats after vitamin E starvation as compared to controls [Tiidus and Houston 1993]. Even when they combined vitamin E deprivation with training, they could not find any additional adaptation in the antioxidant enzymes superoxide dismutase (SOD), catalase (CAT), or glutathione peroxidase (GPX), and their muscle or liver activities [Tiidus & Houston 1994]. No data were provided as to the corresponding changes in vitamin Q content.

Intense Muscle Exercise and Radicals— Oxygen-Burst-Like Conditions?

It is well established in molecular cardiology that the reperfusion injury is a combination of an elevated electron pressure, maximal accumulation of reduced metabolites including lactic acid (lactate), and the "explosive" availability of molecular oxygen (O_2) [Hatori et al. 1989; Braunwald and Kloner 1985; Wolbarsht and Fridovich 1989; Karlsson et al. 1973; Karlsson, Lobstein, et al. 1974]. These biochemical and physiological conditions are also present in skeletal muscle with experimental ischemia, circulatory shock, and arterial balloon pumping support [Karlsson, Lobstein, et al. 1974; Choudhury et al. 1991].

The question is whether intense muscular exercise under normal conditions (i.e., a free blood flow, maximal capillary dilation, and availability of molecular oxygen), in combination with elevated muscle lactates, provides conditions that will promote radical formation, peroxidation, and subsequent muscle trauma.

In addition to the biochemical and physiological conditions (quality aspects), the time function (quantity aspects) should be considered. Short-time maximal exercise is limited by muscle and blood lactate accumulation [Karlsson and Saltin 1971], whereas endurance-type exercise is limited by heat accumulation or fuel exhaustion [Karlsson et al. 1972; Karlsson and Saltin 1971]. Recent experimental data in healthy humans from the August Krogh Institute in Copenhagen, Denmark, indicate that the conditions present during endurance exercise promote more radical injury than short-time maximal exercise does [B. Saltin, personal communication].

Endurance-type exercise is relatively more confined to histochemical trauma in contracting muscles than strength-type sports and training, as has been shown in sport epidemiology studies [Johansson 1987; Sjöström, Johansson, and Lorentzon 1987]. Still, short-time maximal exercise, with its similarity to the oxygen burst injury concept, cannot be ruled out as a significant factor to trigger free radical species production and subsequent biochemical lesions and histochemical trauma.

Muscle Exercise, Purine Metabolism, and Radicals

Heart muscle ischemia, as in infarction, has been shown to trigger radical formation. It has been shown experimentally that this is the result of xanthine oxidation by the enzyme xanthine oxidase (XO), whereby hypoxanthine donates an electron to molecular oxygen (O_2) (see figure 6.5b). The enzyme xanthine oxidase is not present in the healthy heart muscle but appears in relation to ischemia as the result of conversion of the enzyme xanthine dehydrogenase (XDH) [McCord 1985].

In a series of studies, exercise physiologists from the Karolinska Institute in Stockholm, Sweden, studied purine metabolism in healthy humans and with different exercise stress tests. They have not been able to link purine metabolism on the one hand to acute muscle creatine kinase (CK) release to plasma and subsequent CKemia or muscle trauma on the other [Hellsten-Westing, Balsom, Norman, and Sjödin 1993].

Increased plasma hypoxanthine levels originating in the contracting muscles have been demonstrated [Hellsten-Westing et al. 1994; Balsom et al. 1992; Hellsten-Westing, Norman, Balsom, and Sjödin 1993]. In one particular study, running tests of 100 m to 42,000 m (the latter distance being a marathon race) were examined. Plasma hypoxanthine was elevated after all runs. No acute CKemia was present, but it was found after 24 hours following the 5,000-meter and marathon runs [Hellsten-Westing, Sollevi, and Sjödin 1991]. On an individual level, however, no relationship was present between plasma hypoxanthine and plasma CK levels. Most probably this delayed CKemia represented a *de novo* synthesis of the CK enzyme protein. The delayed leakage could also be due to a residue status in the contracting muscles, which remained unrepaired.

Endothelial Tissue

In physiology, from a regulatory point of view, weight and volume (i.e., quantity) can be equally important or even more important than quality. This is relevant for the largest tissue in normal and adult humans—skeletal muscle—which amounts to 20 to 30 kg. In obese people, however, adipose tissue might be larger and relatively more important for blood pressure regulation and insulin secretion.

These considerations are relevant to the largest secretory tissue in man—the endothelial tissue.

Anatomic Aspects of the Endothelium

The endothelial tissue is theoretically a section of the epithelium. It originates from the cell structures lining the embryo. Another cavity lining in the body, originating from the surface during development, is the mesothelium.

The endothelium lining covers the interior of the vessels of the circulatory system. It is in continuous physical contact with the blood, another major tissue of the body. The endothelium lining consists of thin, leaf-shaped cells, continuous with one another at their margins. The continuous nature of the lining is a guarantee that the circulating fluid never touches the interstitial liquid and cells outside the vessels. Most of the plasma contents can, however, freely interchange through the thin endothelial membranes of the capillary network.

Constriction of smooth muscles of the peripheral vessels of the circulatory system is the main limiting factor for the circulatory fluid to escape central circulation and to enter and fill up the volume of the periphery and the capacitance vessels [Holmgren and Åstrand 1966]. The volume of the peripheral circulatory system surpasses several-fold that of the central circulatory volume. An uncontrolled exploitation of this volume should lead to markedly reduced or no venous return, central circulatory hypotension, and collapse; most probably a lethal circulatory shock would develop.

Nitric Oxide Formation and Activity

The endothelial tissue produces an intercellular messenger or signal substance, which has the ability to relax smooth muscle tissue. Biologically, the physical nature of this signal substance is unique. It is an inorganic gas—nitric oxide (NO) (see chapter 2). Moreover, it is electrically neutral, as are all molecules, but in the normal stage it possesses an unpaired electron in its outermost electron orbit; by definition it is a radical species and has some similarities with singlet oxygen ($^1O_2$), for example. As a radical, it represents an example of a nitrogen-centered radical. Formally, it should therefore have the symbol NO·.

NO· acts via formation of cyclic GMP (cGMP). This intracellular messenger or signal substance is produced by the guanylate cyclase system, which is activated by NO·, among other things. Other radical species, such as cumene hydroperoxide and hydroxynonenal, which result from lipid peroxidation, have similar effects [Grune et al. 1993; Martinez et al. 1994]. Of special interest is one particular essential fatty acid belonging to the omega-3 series (vitamin F_1), eicosapentenoic acid (EPA) (see table 2.3), which is a significant stimulant [Furuni and Sugihara 1994].

There is a striking interplay between the two gases oxygen and nitric oxide. The superoxide radical $O_2^{\cdot}$ can destroy $NO^{\cdot}$ and vice versa. The physiological implications of these mechanisms have rapidly expanded, implicating $NO^{\cdot}$ in the regulation of local muscle capillary network dilation and oxygen supply, blood pressure regulation, controlling platelet adhesion, and neutrophil aggregation.

The endothelium formation of the vasoactive radical $NO^{\cdot}$ is referred to as the *endothelium-derived relaxing factor* (EDRF).

Nitric oxide formation is dependent on the amino acid L-arginin as a precursor. Breakdown of L-arginin could be the result of

- circulatory stress as with muscular exercise,
- metabolic pathways in blood platelets,
- release of intracellular metabolites such as ATP and ADP,
- vasoactive hormones such as acetylcholine, thrombin, histamin, and leukotrienes, and
- pharmaceutical products such as nitroglycerine and nitroprusside.

Platelets are a form of blood cell aggregations and assemblies normally present in the circulatory system. Their role in EDRF is dependent on the thrombocyte activity. Under pathological conditions they aggregate in large assemblies, an action referred to as *thrombosis* or *clotting*. EDRF is frequently referred to as *the endogenous nitrate*, with physiological effects comparable to the drug nitroglycerine.

Some speculations about the phylogenetic background and development of this signal system have been provoked by the following facts:

1. $NO^{\cdot}$ is a free radical species, a signal transduction pathway not causing tissue injury, and unique for the endothelium tissue,
2. Endothelium tissue has its origin in the epithelium,
3. $NO^{\cdot}$ is, and has been, a common pollutant (perhaps more prevalent in the early evolution of life on Earth).

The nature of these speculations will not be commented upon here. Such speculations are also present for the origin of mitochondria as subcellular units and the antioxidant scavenging system in general.

EDRF and Ergoreceptor Activity

Muscle lactate formation and accumulation are due to increased muscle activity and recruitment of FT muscle fibers. [Karlsson 1971; Karlsson 1979]. This metabolic condition is one stimulatory feature of the physiologically and pharmacologically defined ergoreceptors, activating the central nervous system (CNS) and the vasomotor units in the elongated brain [Shepherd et al. 1981; Mitchell 1985].

The activation of the CNS and efferent activity to the heart will increase both heart muscle contractility and heart rate, causing an increased cardiac output. Simultaneously, circulatory vasoconstriction will be initiated in resting and semi-active muscles, thereby directing the increased cardiac output and the subsequent blood flow to the contracting muscles that need oxygen.

EDRF Blocking Factors and Systems

EDRF activity is depressed in the presence of other free radicals and by other physiological conditions [Bassenge 1992; Gryglewski, Palmer, and Moncada 1986; Kim, Chen, and Gillis 1992; Mugge et al. 1991]. These regulatory means are to a large extent focused on NO· metabolism.

The reactivity of NO· varies in the physiological milieu depending on the surrounding conditions. The reactivity at diluted concentrations is relatively slow. The reactivity is drastically reduced by either nitric oxide's reaction with molecular oxygen (O_2) to form nitrogen dioxide (NO_2) or with the superoxide radical (O_2^-) to form the negatively charged ion (anion) peroxynitrite (ONOO-). The reactivity is also decreased as the result of another metabolite and adduct of molecular oxygen and NO·, the radical species nitrosyldioxyl (peroxyl nitrite radical)— ONOO·.

Another physiological feature of significance for the fate of NO· is the surrounding pH. At neutral and alkaline pH levels, NO· is masked in the form of other, more stable radicals, which markedly reduces the transducer effect and EDRF [Bassenge 1992].

Taken together, a great number of reactions in which molecular oxygen is directly or indirectly involved down-regulate the presence of the main EDRF factor, nitric oxide.

White Blood Cells and Radical Formation

From a histological point of view, white blood cells (leukocytes) are named and defined as described in figure 6.6. From a functional point of view, they are confined to the immune system and are then referred to as

- T lymphocytes,
- B lymphocytes, and
- phagocytes.

White blood cells all originate from a single type of stem cell in the bone marrow. During cell differentiation they develop as macrophages and lymphocytes.

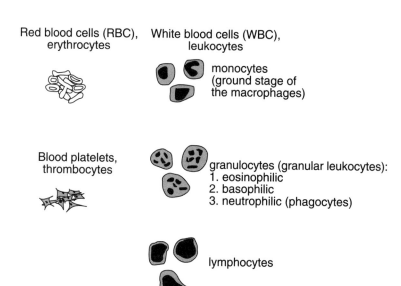

The different cell systems in blood:
1. red blood cells
2. white blood cells
3. blood platelets

Red blood cells (RBC),   White blood cells (WBC),
erythrocytes   leukocytes

monocytes
(ground stage of
the macrophages)

Blood platelets,
thrombocytes

granulocytes (granular leukocytes):
1. eosinophilic
2. basophilic
3. neutrophilic (phagocytes)

lymphocytes

Figure 6.6   The blood is one of the major tissues in our body. Blood tissue contains three cell types: red and white blood cells and thrombocytes. When the cells are excluded, plasma or serum remains (serum is free from coagulation factors and proteins).

The body's immune system and the ability to resist and defeat foreign biological material as viruses, bacteria, and fungi depend on these three cell systems. Even transplanted organs can be the target for attacks from these leukocyte varieties in the rejection process. These "undesired" processes can be reduced or eliminated by immunosuppressive drugs such as cyclosporine.

Bactericidal Activity

The leukocyte subgroup polymorphonuclear neutrophil granulocytes (PMN) is attracted by the microorganisms and foreign cell material and constituents. This will cause activation of their chemotaxis property. They will then start to migrate toward the origin of the attraction. The physical part of this migration is the responsibility of the proteins actin and myosin, the same proteins responsible for muscle contraction.

At the site of confrontation with the foreign material, PMN starts to produce free reactive radicals and bombard the target material. Free radical species are formed in these processes by PMN for two purposes:

- Extracellular formation to injure, weaken, and possibly kill the microorganism to prepare for phagocytosis
- Intracellular formation after phagocytosis to finally kill (if necessary) and decompose the microorganism.

The extracellular radical formation occurs in the outer section of the cell wall or at the cell surface. The intracellular radical formation is organized in conjunction with the phagocytosis *(invagination)* and the corresponding transient granules *(phagosomes),* which generate the superoxide radical (O_2^-) and hydrogen peroxide (H_2O_2). They then trigger formation of the reactive hydroxyl radical (OH') as well as other carbon or nitrogen-centered radicals, such as peroxyl nitrite (ONOO'). H_2O_2 plus the chloride anion (Cl-) form another reactive radical species hypochlorous acid (HOCl) and its anion, the hypochlorite ion (OCl-). These processes, in which reactive oxygen species (ROS) are formed, are frequently referred to as the *respiratory burst* of PMN.

These PMN-based processes are in themselves bactericidal. But they are combined with other granule products and activities such as lysosomes, iron chelation (which deprives the microorganisms of iron-dependent metabolization), cationic proteins, and acidity. This combination will add further metabolic stress to the invading organisms, weaken them, and prevent them from further aggressive attacks. This afflicted stress situation will also provide optimal strategic positions for the final PMN attack, takeover, and phagocytosis. This is the general strategic plan for a successful intervention by the immune system.

It could be argued that nutraceutically elevated plasma and tissue antioxidant levels and activity could impair the PMN respiratory burst and subsequently reduce the bactericidal effect. This possibility of nutratherapy has not been fully examined; therefore, it cannot be ruled out as a potential side effect [Halliwell and Gutteridge 1989].

Inflamed Joints

Joint diseases such as idiopathic (primary) or acquired (secondary) arthrosis frequently combine with leukocyte invasion and inflammatory processes, and arthritis develops. Acquired arthrosis, such as from physical trauma (impact injury), congenital diseases (e.g., metabolic disorders, calcium deposits, and rheumatoid arthritis), and environmental changes or pollution, can be related to biochemical events, indicating cell trauma. These events include increased plasma levels of peroxidized

PUFA, depolymerized and fragmented glucosaminoglucans (muco-polysaccharides), and other constituents of the connective tissue, including hyaluronic acid, phospholipase A_2 activation, and urinary excretion of glucosaminoglucans.

In the engaged joint, unlike in healthy joints, phospholipase A_2 can be determined [Grootveld et al. 1991; Pruzanski et al. 1985]. Moreover, peroxidized PUFA [Rowley et al. 1984; Merry et al. 1991] and fragmentized glucosaminoglucans are present, which also differentiate the inflamed from the healthy joint [Ng et al. 1992; Grootveld et al. 1991]. Treatment with scavengers and inflammatory drugs reduce these changes, which indirectly supports the notion that free radical production takes place in inflamed rather than in healthy joints [McCord 1974; Franson and Rosenthal 1985; Blake et al. 1989; Halliwell, Gutteridge, and Blake 1985].

Miscellaneous

It has been demonstrated that free radical species can act as second messengers [el-Hage and Singh 1990]. Genes, which code for and produce the different proteins involved, are activated by specific nuclear proteins. For this process there are natural inhibitors, which can prevent the translocation of these nuclear proteins from the cytosol to the nucleus. The interplay between the nuclear protein and these inhibitors is the biological basis for maintaining balance. Free radical species have the ability to inactivate this inhibitor mechanism in an indirect activation process. This will lead to increased gene coding, protein synthesis of, for example, phospholipase A_2, and a subsequently aggravated inflammatory activity [el-Hage and Singh 1990]. This system is most probably general in nature and has been proven clinically relevant in relation to inflammatory processes [Williams 1989].

Gene activation by these mechanisms is theoretically applicable to but not experimentally proven for other genes and code systems, such as those that induce adaptive responses in relation to muscular exercise and those responsible for "training effects" or "supercompensation" [Williams 1989]. Most probably, the DNA material in the nucleus and in the mitochondria is sensitive to particular radicals that have specific responses on different genes or gene combinations. Such radical activity on the genome will then generate, for example, actomyosin synthesis and hypertrophy with respect to strength-type training. It will also form certain radical species or increase mitochondrial activity as the response to endurance-type exercise and the result of feedback from the corresponding specific radicals.

Concept of Overuse Injury

The threat of peroxidation injury is confined to muscular exercise. Peroxidation can take place in the contracting muscle and elsewhere (e.g., in engaged connective tissue compartments, the heart, and leukocyte and erythrocyte membranes). There is debate about whether this peroxidation is a result of the combination of normal radical formation with an impaired antioxidant defense or whether it is related to an excessive and pathological radical formation.

The result of these events could be

- muscle inflammation;
- inflammation of connective tissues and related organs, including bursitis and tendinitis;
- rupture of leukocyte cell walls, reduced number of leukocytes, a subsequent reduced immune activity, and increased susceptibility to infectious diseases; and
- hemolysis, reduced arterial blood oxygen content, and oxygen transport capacity (i.e., "runners anemia").

Any of these consequences could result in reduced physical performance directly or indirectly in elite athletes and those enrolled in daily fitness programs. Inflammatory processes could also be the source of everything from discomfort to crippling pain, which occurs as frequently in elite sport as it does in fitness exercise programs.

Summary

In the cell, free radical species are formed by many different mechanisms, both physiological and pathological. Injured cells, ruptured organelles, or the influence of toxic compounds are also sites or triggering factors for radical formation.

Muscle metabolism at rest and during exercise means close regulation of molecular oxygen availability in relation to the need. The muscle is kept at the reduced side. During muscle activity, venous pO_2 is maintained at a low level, indicating the same general goal.

Molecular oxygen is a major source for free radical species generation during muscular exercise. It is estimated that anything between 3% and 15% of the oxygen consumption could pass through an oxygen-centered free radical species or radical-promoting metabolites as hydrogen peroxide.

Crucial and so far uninvestigated questions remain: Is intense exercise, with its maximal tissue lactate accumulation, indicative of an elevated electron pressure? Are the conditions comparable with ischemia and oxygen availability following reperfusion? Can these oxygen-burst-like conditions provoke histochemical trauma to the contracting skeletal muscle in healthy humans?

The endothelial tissue is the largest "secretory organ" in man, providing formation and release of the messenger substance nitric oxide or, formally, the nitric oxide radical (NO·). Nitric oxide exerts a relaxing role on smooth muscle tissue, which causes the blood pressure to extend into the capillary network, resulting in dilation.

The NO· exists in an equilibrium with oxygen-centered radicals. With ample availability of molecular oxygen in the muscle, sufficient amounts of oxygen-centered radical species will be formed to maintain low smooth muscle relaxation and capillary dilation, which is a prerequisite to maintaining a physiologically adequate central arterial blood pressure. An uncontrolled nitric oxide activity could cause a vicious cycle of vasodilation, reduced "afterload," hypotension, and possibly a circulatory shock.

With muscle activity and the subsequent oxygen consumption, the nitric oxide radical activity will be one of many features allowing capillary dilation and local ("micro") blood flow. Increased systemic blood flow, including cardiac output and pulmonary oxygen uptake, will be the result of an increased ergoreceptor activity subsequent to accumulation of metabolites from the fermentative processes.

Leukocytes (white blood cells) are normally inactive and present in the circulatory and the lymphatic systems. They get activated by foreign biological material such as microorganisms and transplanted tissues. This activation means both an active physical movement versus the origin of the stimuli—migration—and an activation of metabolic processes, whereby free reactive radicals are produced. These reactive species attack the microorganisms. This bactericidal activity by the leukocyte-based radical formation is a cornerstone in the immune system of humans.

The radical balance involved in capillary network dilation regulation and the aseptic function of leucocyte radical formation could be interpreted as radicals in service of life. Thus they support our well-being. In this context, it must also be mentioned that radicals are a significant factor in cellular synthesis as they provide the growing cell with important by-products and signal substances.

Radical formation has been associated with inflammatory processes in joints and connective tissue. The triggering mechanisms can vary and in acquired arthritis, for example, responsible triggers can be anything from physical trauma (i.e., impact injury) to metabolic features (i.e., overuse injury). In rheumatoid arthritis, an autoimmune disorder has been suggested as an etiology sequence.

It has been suggested that radical formation and specific radicals act as signals in ultra- and extracellular communication. Thus they serve as a second messenger, besides hormones, that can trigger the DNA coding and protein synthesis responsible for enzyme formation (e.g., phospholipase A_2). Radical formation is an aggressive component in the inflammatory process and might cause release of precursors for aggressive eicosanoid hormone synthesis. However, radical formation will also trigger genes responsible for enzyme protein synthesis related to supercompensation and adaptation to muscle exercise with physical training.

7

Exercising Muscle and Radical Formation

The existence of free radicals in biology has been disputed as recently as in the 1980s. I have been questioned several times by highly qualified oxygen biochemists about the relevance of radicals in healthy people who exercise. Most researchers agree, however, that radical formation does exist in the biological arena and that it can be related etiologically to such disorders as cardiovascular diseases, cancer, and diabetes, and that it may also promote the onset of the aging process although not necessarily shorten the life span [Kanter 1994]. Life span depends more upon the atopose gene and "programmed" cell death.

Biological Evidence of Free Radicals

Several reviews have covered the field of free radicals in medicine generally [Halliwell 1987] and in sports medicine in particular [Jenkins 1993; Kanter 1994].

Radical, or Radical Probe Detection

In biochemistry, electron spin resonance (ESR) or nuclear magnetic resonance spectroscopy (NMRS) has been used for a long time to study radical metabolism. The same technique has also been applied to *in situ* animal experiments. Two approaches have been used:

- Direct observation of these short-lived species
- Indirect observation by trapping (spintrapping)

In the latter method, the radical species is allowed to react with a certain foreign compound (xenobiotics) to produce a more stable product or probe.

Another approach is to register the energy content and release from radicals or otherwise excited species. In animal exercise physiology, Jenkins has successfully applied chemiluminescence (CML) to detect single photons released from singlet oxygen [Boveris et al. 1980]. When the energized oxygen molecule—singlet oxygen ($^1O_2$)—allows its unpaired electron to return to its original orbit, energy is released. The energy released can be detected as light, as a photon will be emitted.

Fingerprinting in Humans

In humans, neither of these methods is applicable to any major extent. Indirect methods, referred to as *fingerprinting*, must be applied. According to this approach, a radical is inferred from the molecular nature of the damage it causes to biological molecules.

If oxidative stress is great enough to overcome the antioxidant defense, peroxidation will occur. The reactive radical species can damage practically every component of the cell, including proteins, nucleic acids in DNA and RNA, lipids in general, and the structural lipids of polyunsaturated fatty acids in particular. These damaged molecules or the products resulting from their breakdown are the fingerprints. These techniques and the corresponding results, as indirect methods to mark for postulated radical-induced muscle trauma, are inconclusive and debatable.

Protein Damage and Fingerprints

Radical-induced trauma to proteins can be described by the appearance of modified amino acids in the form of protein carbonyls [Stadtman and Oliver 1991; Levine et al. 1990].

Exercise in humans is synonymous with a negative nitrogen balance (i.e., a net breakdown of protein and deamination) [Goldspink 1991].

However, the physiological background to this degradation is not well defined. Scientists do not agree about the nature of protein degradation with exercise in healthy humans, although radical formation and protein damage may explain it.

Exercise-induced myalgia and creatine kinase (CK) release to plasma (CKemia) related to radical-induced trauma has been documented in pathological conditions as well as in gene mutations [Hatae et al. 1991].

Nucleic Acid and Fingerprints

Unrepaired DNA and RNA damage can form and can be estimated by means of 8-hydroxy-2'-deoxyguanosine determination (8-HOG) in urine [Shigenaga, Gimeno, and Ames 1989]. Ames and colleagues reported in 1988 that DNA damage in different mammal species increased with the average metabolic rate [Adelman et al. 1988]. They found that the metabolic rate and the subsequent DNA damage was inversely related to body size of the species.

Urine excretion of nucleosides following physical exercise has been taken as evidence of nucleic acid damage [Alessio 1993; Alessio et al. 1990]. Thus, urine content of 8-HOG has been found to increase after a marathon race (see figure 7.1).

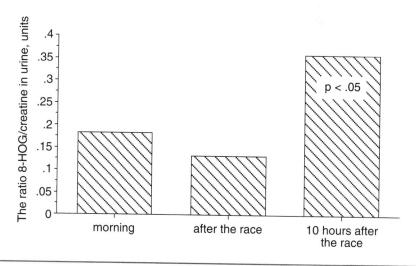

Figure 7.1   Urine excretion of 8-hydroxy-2'-deoxyguanosine (8-HOG) as a marker of DNA damage and production of oxidized nucleosides [Alessio et al. 1990; Alessio 1993].

Lipid Peroxidation and Fingerprints

Lipid peroxidation can be determined by means of plasma analysis of thiobarbituric acid reactive substances (TBARS) [Halliwell and Grootweld 1987]. One of these TBARS is malondialdehyde (MDA). This metabolite of lipid peroxidation is frequently used as a fingerprint marker in clinical work to describe the extent of peroxidized lipids [Noberasco et al. 1991]. It is more specific and, consequently, more reliable than determination of all TBARS present.

People who exercise are reported to have elevated TBARS in plasma following exhaustive exercise [Kanter et al. 1988]. In a study with a cycle ergometer exercise protocol leading to exhaustion, conflicting results were obtained [Sahlin, Ekberg, and Cizinsky 1991]. It was possible to describe a positive femoral venous-arterial difference over the contracting leg muscles for hypoxanthine. Plasma MDA was, however, not at all detectable. Thus, the validity of either the relevance of MDA determination as a fingerprint method or the actual study protocol is questionable [Sahlin, Ekberg, and Cizinsky 1991].

TBARS are also present in humans in urine excreted after eccentric exercise [Meydani et al. 1993].

A different experimental approach is to examine for hydrocarbon gases in expired air [Riely and Cohen 1974]. The different alkanes—ethane and pentane—are the results of omega-3 and omega-6 fatty acid peroxidation, respectively [Pryor 1993].

When lipid peroxides dissociate, the hydrocarbons ethane and pentane are produced. However, these alkanes are indirect markers of lipid peroxidation, and the formation is not stoichiometrically related to the magnitude of lipid peroxidation. Moreover, both alkanes are metabolized, but pentane more rapidly than ethane. Pentane formation may be the result of cell-membrane-allocated lipids and their peroxidation [Pincemail, Deby, and Dethier 1987; Riely and Cohen 1974].

It has been shown experimentally that exhaled pentane as a fingerprint of lipid peroxidation increases linearly with relative exercise intensity; the absolute intensity increases as a fraction of maximal pulmonary oxygen uptake ($\dot{V}O_2$max) (see figure 7.2a) [Dillard et al. 1978]. It has also been shown that a nutraceutical supplement program with vitamin E almost abolished pentane levels in the expired breath (see figure 7.2b) [Pincemail, Deby, and Dethier 1987].

Lipid peroxidation can be traced by other means, such as with analysis of conjugated dienes, hydroperoxides, and short-chain hydrocarbons. Thus, lipid-conjugated dienes in skeletal muscle might increase with exercise in humans [Meydani et al. 1993]. Hydrocarbons are expired in humans after radical-induced trauma as the result of smoking, but so far, to my knowledge, not with exercise [Duthie 1993].

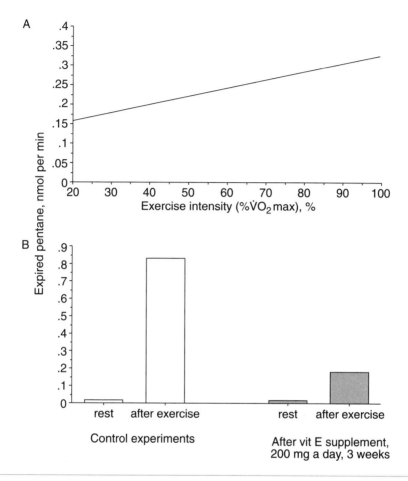

Figure 7.2 (a) Peroxidized lipids and their further metabolism include production of hydrocarbons: ethane from omega-3 and pentane from omega-6 fatty acids. With a diet typical of the rural western hemisphere, the omega-6 fatty acids dominate in membranes and lipoproteins [Alessio 1993; Dillard et al. 1978]. Pentane increases in the expired air linearly to the relative exercise intensity, i.e., the actual pulmonary oxygen uptake ($\dot{V}O_2$), in percent, of the individual maximal pulmonary oxygen uptake ($\dot{V}O_2max$) (($\dot{V}O_2$) x ($\dot{V}O_2max$)$^{-1}$ x 100, %($\dot{V}O_2max$). (b) A nutraceutical treatment program (nutratherapy) with vitamin E for 3 weeks (200 mg a day) markedly suppressed expiration of pentane following a standardized exercise test [Pincemail, Deby, and Dethier 1987].

Summary

Techniques that directly demonstrate reactive radical species in humans are lacking. Infusion of xenobiotics as probes with well-defined reactions with radicals in combination with nuclear magnetic resonance spectroscopy represent a new but so far unexplored avenue.

In humans, analysis of either stable radicals or reaction products of reactive radicals—fingerprint studies—have been applicable. Lipids, especially unsaturated fatty acids, and formation of ethane (from omega-3 fatty acids) and pentane (from omega-6 fatty acids) as a result of peroxide degradation is one approach. These hydrocarbon (alkane) gases can be detected in expired air. With muscle exercise, alkane exhalation increases with intensity. The production of hydrocarbons can be reduced by long-term nutratherapy with antioxidants.

Studies have been done that indicate damage to nucleic acids in DNA and RNA molecules and even protein damage as the result of reactions with reactive radical species.

8

Nutrients as Antioxidants and Their Food Sources

The term *vitamin* and its application has caused debate through the years. The *Geigy Scientific Tables* defines vitamins as "essential food components that are organic in nature but which . . . are needed only in small amounts." [Geigy, 1986]. This definition, however, could exclude vitamins C (ascorbate) and B_3 (niacin or nicotinic acid) because they are consumed in dosages larger than "small amounts."

The *Geigy* encyclopedia, like most of the literature in the field, separates vitamins into two groups: fat-soluble (lipophilic) and water-soluble (hydrophilic). Antioxidant vitamins are present in both groups. Coenzymes or their constituents, as well as "cell spare parts," dominate among the lipophilic vitamins.

With muscle exercise, radical formation dominates in different subcellular lipoidic compartments. The major radical formation control mechanisms ("the first defense line") are therefore located in these membranes. By means of the antioxidant vitamin Q-E-C cycle, the radicals, or rather the corresponding electrons, are transferred to their hydrophilic antioxidant vitamin counterparts. Ultimately, the antioxidant enzyme systems SOD-CAT and GPX-GRD recover the electrons for the respiratory chain.

Some aspects of food antioxidant nutrients are now presented to show the extent to which food sources and the quality of these sources are significant.

Food Sources for Antioxidant Nutrients

All cells, whether animal or plant, contain structural lipids in the cell wall and other membrane structures. These structural lipids are relatively richer in polyunsaturated fatty acids than, for example, triglyceride (TG) deposits in adipose cells. Structural lipids and adipose-based lipids can store lipid-soluble compounds. Adipose cells or tissues do not contain as high amounts of lipid-soluble vitamins as structural lipids do. The reason for this is not only smaller amounts of the vulnerable PUFA. Additionally, adipose-tissue-allocated triglycerides are stored there as an energy fuel source and are relatively inert metabolically until mobilized and lipolyzed. The condition of metabolic inertia, by definition, means less exposure to reactive radical species and, therefore, less need for antioxidant protection.

Regulation of Vitamin E Content

Vitamin E content in different food sources varies from zero (0) up to 30 to 50 mg per kg in, for example, sunflower seeds or palm oil. Simultaneously, fat content expressed as triglycerides can reach values in excess of 40% in sunflower seeds, nuts, meat products such as bacon, or refined oil products such as soybeans, corn, fish liver, fish muscles, and fish meat oils. Correlation matrix analyses have revealed that vitamin E content increases with the amount of PUFA in both absolute and relative terms (see table 8.1).

For the different food sources used in table 8.1, equally distributed among fruits, vegetables, nuts, grains, oils, meat, and poultry, vitamin E content increases almost linearly with lipid content expressed as TG (see figure 8.1a).

In relation to the PUFA content, the increase in vitamin E appears to level off, which might indicate "saturation-like" conditions with PUFA values in excess of 200 to 300 mg per kg, providing a satisfactory vitamin E availability (see figure 8.1b). An alternative explanation could be an increased turnover of vitamin E and a subsequent state of deficiency. This deficiency has been suggested as the reason for the relatively lower

Table 8.1   Correlation Matrix for Vitamin E, Triglyceride (TG), and PUFA as the Relative Fraction of TG [PUFA] × [TG]$^{-1}$ × 100, or PUFA/TG × 100

	Vitamin E	TG	PUFA	PUFA/TG × 100
Vitamin E	1			
TG	.52	1		
PUFA	.58	.71	1	
PUFA/TG × 100	.58	.05	.62	1

vitamin E content in the lung and in the liver as compared to other tissues [Burton and Ingold 1989].

When the PUFA content is expressed as a fraction of the total lipids, a linear relationship seems to be present (see figure 8.1c).

It is suggested that the vitamin E content is relatively small in membranes and on the order of 0.1 to 1 per 100 moles of phospholipids (0.1 to 1.0 mol percent) [McMurchie and McIntosh 1986]. It is also suggested that the relationship between vitamin E and PUFA in biological conditions is 1 molecule per 1,000 PUFA molecules, or 0.1 mol percent [Tappel 1980]. It has been concluded that one vitamin E molecule protects 700 to 1,000 PUFA molecules.

Regulation of Vitamin Q Content

There is much less literature on vitamin Q as a nutrient and its food sources than there is on vitamin E. The natural occurrence of vitamin Q was described in a classic scientific report by Frederick Crane [Lester and Crane 1959]. Crane and associates, two years earlier, reported on the orange-colored molecule in beef heart mitochondria [Crane et al. 1957].They analyzed for the different vitamin Q species identified in biological conditions: Q_6 to Q_{10}. In animal, bird, and eatable amphibian species, it was found that heart muscle contained two to three times more vitamin Q than skeletal muscle contained. Lamb heart, for example, was found to contain 160 mg per kg. Lobster and shrimp seemed to lack this compound, whereas some insects had measurable amounts. Algae and higher plants had values on the order of 10 to 20 mg per kg.

The content of vitamin Q in different food sources varies from almost nothing to about 100 mg per kg (in soybean oil) (see figure 8.2a) [Kamei et al. 1986; Ramasarma 1985]. There is evidently no consistent pattern with regard to the corresponding vitamin E content (see figure 8.2b).

In addition to these differences in vitamin contents in different food sources, there are seasonal variations. In winter, both in Europe and North America, vegetables are shipped in from the southern parts of the

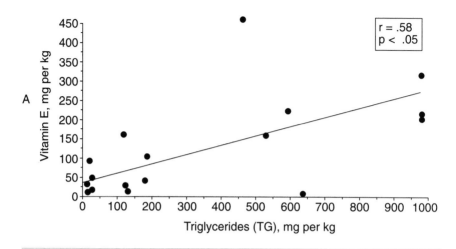

Figure 8.1a   The relationship between vitamin E content in different food sources and the corresponding TG content. Data are collected from different handbooks (e.g., *Geigy Scientific Tables)* [Geigy, 1986].

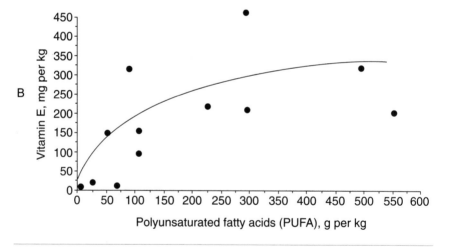

Figure 8.1b   The corresponding relationship between the vitamin E content and the contents of PUFA.

two continents. They can also be artificially grown. Preliminary data have indicated the nutritional significance of these factors.

The Swedish National Food Administration has computed the daily intake of vitamin Q as 2 to 20 mg per day. The basis for this estimate is a Japanese study in the vitamin Q content of different food sources [Kamei et al. 1986]. Although this estimate is very likely representative of the major urban areas of the western hemisphere, regional differ-

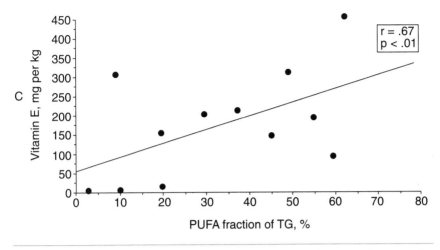

Figure 8.1c   The vitamin E content in the food sources above in relation to the PUFA fraction of the total TG content.

ences might exist, especially in coastal areas where fish is one of the basic protein sources.

The extent to which a well-balanced, mixed diet fulfills our vitamin Q needs is presently unknown [Karlsson et al. 1990a; Bogentoft et al. 1991; Yamagami, Shibata, and Folkers 1976]. Reports have described how food processing and other interventions lower the vitamin Q content in food [Zamora, Hidalap, and Tappel 1991; Huertas et al. 1991]. The ways in which the endogenous production of vitamin Q and absorption by the digestive system can be hampered have also been described [Folkers et al. 1990; Willis et al. 1990; Bélichard, Pruneau, and Zhiri 1993; Hübner et al. 1993]. Taken altogether, it has been claimed that vitamin Q, like some of the PUFA, is a conditionally essential nutrient.

Antioxidant-Related Nutrients

It is recognized that a major protective goal for antioxidants is to avoid radical induced trauma to PUFA, in general, and EFA, in particular. There are two EFA groups or series: the omega-3 and omega-6 fatty acids (see table 8.2). (The third major group of PUFA, the omega-9 series, can be synthesized in sufficient amounts in humans.) The two EFA groups were given the name *vitamin F* in the 1920s, as they were the sixth group of nutrients found to be essential for life (vitamins A through E representing the first to fifth groups) [Food and Drug Administration 1979]. The indices F_1 and F_2 have been added to distinguish between the omega-3 and the omega-6 series, respectively.

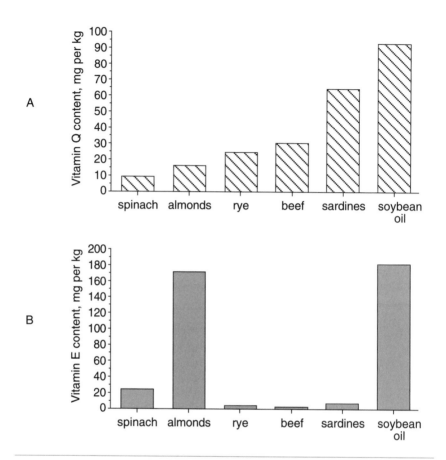

Figure 8.2 (a and b) Contents of vitamins Q (a) and E (b) in different food sources. Data are collected from different handbooks, e.g., *Geigy Scientific Tables* [GEIGY, 1986] and other scientific contributions [Ramasarma, 1985].

The term *essential* is debated with respect to vitamins F_1 and F_2. The first representatives of each series—alpha (α)-linolenic acid (ALA) of the omega-3 series and linoleic acid (LA) of the omega-6 series—are the true essential fatty acids. The other fatty acids are formed by means of desaturase and elongase enzymes from these precursors. However, ALA is especially lacking in the western diet, and the two series are in competitive situations at the site of joint pathway enzymes. This has been the rationale for referring to the remaining fatty acids of the two series as *conditionally essential* nutrients and, consequently, all omega-3 and -6 fatty acids (vitamins F_1 and F_2) as *essential*.

In an intriguing series of reports, Skjervold discussed ethnosocial and ethnonutritional aspects of food intake in humans [Skjervold 1991]. His main points follow:

Table 8.2   Essential Fatty Acids and the Two Series*

Omega-3 fatty acids (Vitamin F_1)	Omega-6 fatty acids (Vitamin F_2)
α-(alpha-) linolenic acid (ALA) (18:3, Ω-3)	linoleic acid (LA) (18:2, Ω-6)
octadecatetrenoic acid (18:4, Ω-3)	γ-(gamma-) linolenic acid (GLA) (18:3, Ω-6)
eicosapentenoic acid (20:4, Ω-3)	dihomo-γ-linolenic acid (DGLA) (20:3, Ω-6)
eicosapentenoic acid (EPA) (20:5, Ω-3)	arachidonic acid (AA) (20:4, Ω-6)
docosapentenoic acid (DPA) (22:5, Ω-3)	adrenic acid (22:4, Ω-6)
docosahexenoic acid (DHA) (22:6, Ω-3)	docosapentoneic acid (22:5, Ω-6)

The Ω-6/Ω-3 fatty acid ratio
the ratio: (Ω-6) × (Ω-3)⁻¹ or Ω-6/Ω-3 fatty acids

Relative EPA
the ratio: (EPA) × (AA)⁻¹ × 100 or EPA/AA × 100

*Clinically applied ratios to describe the relationship

- When humans were separated from the other primates some 6 million years ago, a change in our forefathers'dietary habits took place.
- Plants and plant parts had earlier been important ingredients in our predecessors' food intake [Eaton and Konner 1985]. This diet, however, was different from present vegetable food sources such as commercial cereal products.
- As humans differentiated themselves from the primates, meat became an increasingly dominant part of the diet. Meat sources were originally hunted game. With the Stone Age or just before that, animals were domesticated or "adopted" by man, resulting in our current varieties of dogs, horses, and cattle.
- Meat sources of our Stone Age ancestors had a 4% fat content; our own meat sources are 25% to 30% fat. Their average fat consumption as a portion of the total energy intake is estimated to have been

20% to 25% (E%F); present proportions of fat are generally between 40% and 50%.

- The increased fat content in livestock meat is due to the preferences of urban people and the impact of their demands on how the livestock is raised and fed.

- The relative proportion of vitamin F_1 was originally higher with respect to both the first representative of the series (ALA), and one of its metabolites (EPA). The ratio omega-6 over omega-3 fatty acids (the Ω-6/Ω-3 ratio) averaged 2:1 [Weber 1989; Dyerberg et al. 1978] (see table 8.2).

- The quantitatively and qualitatively different lipid patterns in our livestock, as compared to those in wild game and the livestock of the early urban era, were due to breeding and/or the type of food grain given the animals.

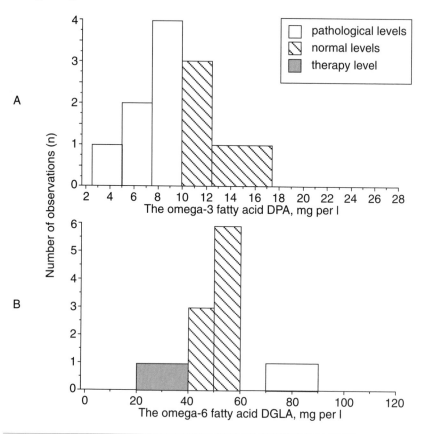

Figure 8.3   (a and b) Plasma values for the omega-3 fatty acid DPA (a) and the omega-6 fatty acid DGLA (b) in a control population with a mixed diet and no nutraceutical treatment.

Table 8.3   Essential Fatty Acids and Vitamins Q and E: Actual Intake and Unofficially Recommended Values for Daily Intake for Average Persons and Elite Athletes (the information assumes a diet representative of the urban western hemisphere)

Nutrient	Intake	Recommended daily allowance (RDA)	Elite athletes' RDA
SFA (g)	55 (25%)	35 (10%)	-
PUFA (g)	50 (15-20%)	70 (20%)	-
Omega-3 fatty acids (g)	0-5 (0-2%)	5-10 (2-3%)	-
ALA	<0.5	1	-
EPA	0.5	3	4-6
DPA	<0.1	0.5	≈1
DHA	0.5	1	2-3
Omega-6 fatty acids (g)	10-30 (3-9%)	15 (5%)	-
LA	-	-	-
GLA	-	-	-
DGLA	-	-	-
AA	26	8	-
Ω-6/Ω-3 ratio	≥10	≈5	1-2
EPA/AA × 100	≈10	>30	>50
Lipophilic antioxidants (mg)			
vitamin Q	2-20	60	100-300
vitamin E	3-15	100-200	300-1,000

Investigations of more recent populations have demonstrated that the Ω-6/Ω-3 ratio is 50 in the urban parts of the western hemisphere, 12 in Japan, and 1 among nomadic Eskimoes [Weber 1989]. Thus, in the western hemisphere, the normal diet favors the omega-6 series (see figure 8.3). The absolute and relative consumption of one of the omega-6 fatty acids—the arachidonic acid (AA)—has been of particular interest (see table 8.2). This fatty acid is the precursor of the most aggressive of the eicosanoids—the prostanoids (prostaglandins) PGE_2 and PGI_2 and the leukotrienes LTC_4 and LTE_4. The formation of these eicosanoids in relation to, for example, inflammatory processes is referred to as the *arachidonic acid cascade*. The omega-3 fatty acids suppress the omega-6 fatty acids in absolute and relative amounts.

Epidemiology has emphasized the deleterious role of arachidonic acid in a number of inflammatory disorders such as rheumatoid arthritis [Horrobin 1986; Leventhal et al. 1993], atopic eczema [Søyland 1993; Søyland et al. 1993; Burton 1989], and asthma [Rocklin et al. 1986]. Clinically, enhanced plasma omega-3 fatty acid and suppressed omega-6 fatty acid values coincide with less inflammatory responses [Leaf and Weber 1988; Burton 1989; Drevon 1992].

In the western hemisphere, the normal diet favors the omega-6 series. The condition of enhanced plasma omega-3 fatty acids values can be obtained by means of dietary changes and/or nutratherapy. Discrepancies exist between the actual intake in man of vitamin F and the recommended values (see table 8.3). These recommendations are not officially recognized as Recommended Daily Allowance (RDA) values by international nutrition authorities.

The exploitation of livestock and the development of agriculture in the last 10,000 to 50,000 years have caused gene drifts. Economical aspects of production and the taste for certain products have been driving forces. Qualitative changes have been made in our major food sources and in their micronutrient contents. Only in recent years have these facts and the corresponding nutritional aspects of our food intake been appreciated outside the scientific community. This has caused a general and political debate ("the Green Movement" during the 1980s and later), which is in line with the expanded general information and knowledge about micronutrients, in general, and essential fatty acids and antioxidants, in particular.

Summary

The term *vitamin* has been used in a broad sense and includes both ubiquinone (vitamin Q) and the essential fatty acids of the Ω-3 and Ω-6 series, or vitamin F (F_1 and F_2).

Food sources vary considerably with respect to their nutritional quality. Vitamin E has been one of the most thoroughly investigated nutrients. The content of fats and especially PUFA seems to be a major determinant for vitamin E content in general. This means, then, that nature has allocated the content of antioxidants in relation to the risk for peroxidation by reactive radical species.

The nutritional value of vitamin Q has been researched less thoroughly than that of vitamin E. The reason may be its relatively late discovery.

Vitamin Q is both an endogenous substance and a micronutrient. The discovery of vitamin Q and the possible disorders related to its deficiency are the result of clinical research that began in the mid-1970s.

Variations in vitamin Q content in different food sources are on the same order of magnitude as vitamin E. The appearance of vitamin Q in food does not mimic that of vitamin E. It is also conceivable that in animal food sources, the type of tissue has bearing on the allocation of vitamins Q and E.

EFA are protected by the antioxidant vitamins Q and E. EFA contain two groups: the omega-3 and omega-6 fatty acids (vitamins F_1 and F_2). The present dietary habits in the urban western hemisphere favor an excessive intake of vitamin F_2 while intake of vitamin F_1 is insufficient. A major source of vitamin F_1 is fatty, ocean fish; a major source of F_2 is beef meat. Improvement of the nutritional vitamin F_1 status suppresses the formation of the aggressive prostanoids originating from the omega-6 fatty acid—the arachidonic acid. Clinically, this EFA consumption pattern and the corresponding plasma values coincide with different undesirable immune reactions in the skin and joints, which can be suppressed by elevated plasma omega-3 fatty acid levels.

Changed plasma EFA values are obtained by dietary changes or nutraceutical treatment (nutratherapy or food supplement programs). A common nutratherapy is, therefore, a combination of vitamins Q, E, and F_1. Fish muscles (fish meat) are rich in vitamin F_1, which is also available as a nutraceutical in the form of a fish oil concentrate.

9

Exercise, Mixed Diets, and Nutratherapy

Muscle exercise can be difficult to describe. Terms such as *heavy* or *maximal* are frequently used to describe muscle exercise without any common consensus about their meaning.

Symptom-limited, or maximal, exercise means that an exercise is generally terminated due to these signs:

1. Discomfort, pain, or exhaustion in the contracting muscles
2. Pain elsewhere in the body directly or indirectly related to the exercise task (e.g., chest pain in effort angina or low-back pain)
3. "Central fatigue" due to psychoneurological changes, dehydration, failing arterial systemic blood pressure, and so on

Muscle Exercise and Its Limitations

Symptom-limited or maximal muscle exercise can last for a few muscle contractions or can occur as dynamic muscle exercise for several hours. Limiting factors vary, however, depending on such features as the muscle volume involved, mode of exercise, and relative exercise intensity [Karlsson 1979]. There are no major differences in limiting factors for exercise between fitness activity participants and elite athletes.

One critical issue is the type of metabolism recruited to supply chemically bound energy to the contractile mechanism, or rather the ATP-ases, represented by the protein conglomerate actomyosin. The energy is ultimately provided by means of ATP. In general terms, three sources of chemical bound energy are available for different metabolic processes:

Energy Source	Type of Metabolism
Local ATP and CP stores	Splitting
Glucose-glycogen	Fermentation
Carbohydrates and fat	Respiration

From a quantitative point of view, the first source—local ATP and CP stores—is insignificant when compared to the amount of ATP formed through fermentation or respiration [Agnevik et al. 1967; Karlsson 1971]. The high-energy phosphate (phosphagen) stores are extremely relevant, though, for short-time maximal exercise performed in sports such as jumping, shotputting, and even the 100-meter dash [Karlsson 1979]. It is estimated that the local phosphagen stores can cover, at most, 5 to 7 seconds of heavy muscular exercise (see figure 6.5b) [di Prampero 1981].

Muscle exercise lasting for more than around 10 seconds is, from a physiological point of view, quantitatively dependent on carbohydrate (CHO) fermentation and respiration, as described in figure 9.1a and b. In relative terms, fermentation is the main ATP source for short-time exhaustive exercise: more than 50% of the energy output for performance times on the order of 40 to 50 seconds, at most. The longer the performance time (more than 40 to 50 s), the more significant respiration will become. Exhaustive exercise lasting 5 to 7 minutes or more is almost exclusively dependent on respiration for energy delivery.

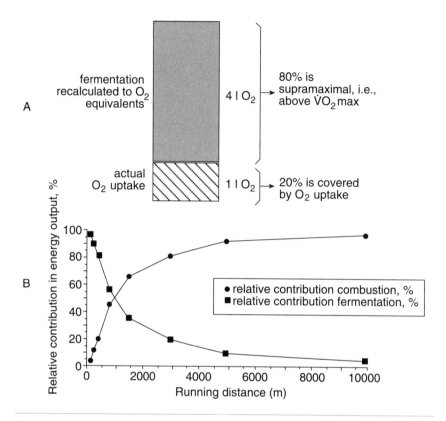

Figure 9.1 (a) The energy output in a 400-meter race according to oxygen uptake determinations during and after a race to compute oxygen deficit and debt conditions. Repeated blood lactate determinations are included to support the oxygen debt data [Agnevik et al. 1967]. (b) The computed relative fractions of fermentation- and respiration-derived energy during different running events, which from a physiological point of view are covered by the processes described in figure 9.2a. Shorter "maximal" performances, including the 100-meter dash, are, to a significant extent, covered by local stores of ATP and CP (the phosphagens) [Karlsson 1971]. For simplicity's sake, this factor is excluded from the fraction calculations.

Muscle Activity and Energy Intake

In the sedentary stage, energy intake for healthy young females and males is around 1,400 kcal (6 MJ) and 1,800 kcal (8 MJ) a day. It increases to about 2,000 kcal (8.5 MJ) and 2,500 kcal (10.5 MJ) with "white collar" activities. The energy requirement falls at a rate of 5% per decade between the ages of 40 and 59 years, 10% between the ages of 60 and 69 years, and another 10% after age 70.

The energy requirement in different sports varies from 2,500 to 5,500 kcal (10.5 to 23.0 MJ) and 3,000 to 8,000 kcal (12.5 to 35 MJ) a day in the two genders. Occasionally, it might increase to 10,000 kcal (42 MJ) and 13,000 kcal (55 MJ) or more, for example, in connection with extremely competitive cross-country (XC) races such as the Vasa Ski Race (85 km) in Sweden. Similar high-energy intakes are also present in ultra-marathon running, bicycle racing (as in the Tour de France), triathlons, and so on.

Endurance sports are characterized by a high gross energy intake. Under these conditions, CHO is a major fuel and the carbohydrate portion of the total energy intake (E%CHO) is increased (see figure 9.2a). As a result, the relative significance of lipids (L) as a fuel (E%L) is decreased (see figure 9.2b). Values less than 25% can be expected with extremely high energy outputs [van Erp-Baart et al. 1989a].

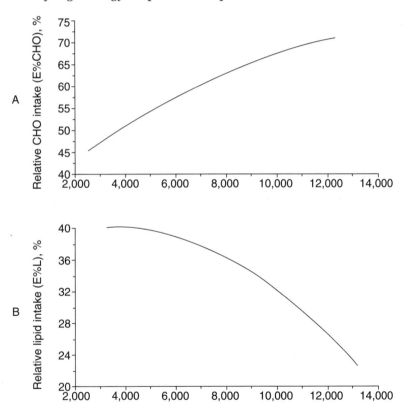

Figure 9.2   (a and b) The computed fraction of carbohydrates (CHO, a) and fat or lipids (L, b) in the food intake of athlete groups with varying training and competition schedules. The CHO and L fractions, respectively, are expressed as a percent of the gross energy intake (E%CHO and E%L).

Relative lipid intakes at such low figures represent an absolute fat intake on the order of 100 to 200 g per day. Assuming 3 hours' training activity with a pulmonary oxygen uptake of about 5 L per minute, it can be estimated that about 40 g of fat are combusted. During the remaining hours of the day, another 50 to 100 g of fat are used. Taken altogether, there is an obvious risk for endurance athletes to have a lipid intake that barely covers the fuel needs. The turnover of structural lipids and, for example, the essential fatty acids is left aside. In sports medicine, this is referred to as the *carbohydrate trap*. In its chronic form, it is an example of malnutrition and an eating disorder.

Energy Intake and "Natural Nutrients"

The vitamin E intake in elite Swedish XC skiers in the early 1990s, as presented in chapter 2, did not exceed 50 mg a day. The plasma vitamin E levels were found to be linearly related to the estimated vitamin E intake (see figure 2.4).

It has been demonstrated in several studies that nutrient intake, in general, increases linearly with the energy intake [van Erp-Baart et al. 1989b; Blix 1965]. According to studies done by the Maastricht group, the computed intakes of critical nutrients such as vitamins A, B_1, B_2, B_6, and C are equal to or above the national guidelines of the Recommended Daily Allowance (RDA) providing the athletes had a nutritious, well-balanced, and mixed diet [van Erp-Baart et al. 1989b]. No blood or plasma data were saved to document a satisfactory plasma level of these nutrients.

Such information has been the rationale for statements banning food supplements or denying any beneficial effect from them. In 1991 the International Olympic Committee (IOC) Medical Commission published a statement in its "Consensus Decision at Lausanne" [Devlin and Williams 1991]. According to this statement, a mixed diet is satisfactory to cover the nutrient needs of elite athletes.

What does it mean to have a nutritious, well-balanced, and mixed diet? As late as 1994, a book from the IOC Medical Commission recommended a standard diet for elite soccer players based on 60% to 65% carbohydrates with regard to energy content (E%CHO), 15% lipid (E%L), and 20% to 25% protein (E%Pr) [Ekblom 1994]. Such a diet is not a balanced diet. It is acceptable for two or at the most four days within the framework of a carbohydrate supplement and muscle glycogen loading program [Karlsson 1979]. It is debatable, however, as to whether such a diet could initiate the vicious cycle in dietary habits represented by the anorectic eating disorder anorexia nervosa.

A second set of endurance exercise studies has been carried out by the Maastricht group, but in this case at the fitness level. They studied females and males engaged in a fitness exercise program that included a marathon race. Mean performance times were 3 hours 52 minutes for the women, and 3 hours 41 minutes for the men. By European standards, the physical performance capacity of these athletes was above average.

The investigators could document clear indices of local muscle trauma similar to those observed as the result of radical-induced injuries. Thus, they found both elevated thigh muscle (*m vastus lateralis*) adenosine [van der Vusse et al. 1989] and a negative relationship between muscle carnitine (vitamin O) and plasma CK activity (i.e., CKemia) (see chapters 6 and 7) [Janssen, Scholte, et al. 1989]. Adenosine is suggested to be a precursor for reactive radical species formation (see chapters 2 and 5, and figure 6.5b for more on reactive radical species). But there are equally strong alternative sources for radical formation and muscle trauma under these conditions.

The CKemia finding should be interpreted as follows:

1. Muscle CK and carnitine depletions were documented
2. Increased plasma CK activity (CKemia) was present
3. Low muscle carnitine coincided with CKemia.

The Maastricht group also reported that "long-distance running is associated with transient minor pathological changes in skeletal muscle" [Kuipers et al. 1989, p. S156]. It could also be shown that a relationship existed between muscle fiber histopathology and the corresponding plasma enzyme increases [Janssen, Wersch, et al. 1989].

Overuse injuries were therefore already present with an exercise intensity corresponding to an ordinary fitness physical conditioning program. The participants had only a mixed diet and no nutratherapy. Overuse injury could also be documented as red and white blood cell hypoplasia and less immune reactivity [Janssen, Wersch, et al. 1989].

The investigators' findings with regard to adenosine might be of special interest. Adenosine is a potent vasodilator in most vascular beds [Berne 1980; Sollevi 1986]. Adenosine causes a relaxation of vascular smooth muscle, which is mediated by specific cell-surface adenosine receptors [Schrader et al. 1977]. As it also contributes to the threat of radical trauma together with other trauma-related metabolites, its washout is crucial for the tissue of adenosine origin.

These findings in participants in fitness programs should be evaluated based on the information presented in chapter 2 (see figure 2.1a), particularly the sport epidemiology study carried out at the Umeå University, Sweden, in elite athletes from different sports [Sjöström,

Johansson, and Lorentzon 1987; Johansson 1987]. The major finding was that muscle fiber histopathology dominated in endurance-type athletes but was rare in athletes involved in strength-type training.

In doctoral studies that I have guided on long-distance runners with lower leg muscle pain (see figure 2.1b and c), it was found that those with a relatively lower percent distribution of capillary-rich, slow-twitch (ST) muscle fibers had more muscle edema and muscle lactate in pain provocation exercise tests [Karlsson and Smith 1984; Wallensten and Eriksson 1984]. One can infer from this finding that those rich in fast-twitch (FT) muscle fibers are less prepared to meet the challenge of reactive radical species formation. In twin studies, a major factor determining individual muscle fiber composition was found to be the genetic heritage [Komi and Karlsson 1979]. It seems reasonable to suggest that this endowment-related trait to conquer reactive radical species could be one of several biological clues to recruit or select potential endurance athletes from within a population.

Elite Sport Activity and Antioxidant Nutratherapy

The United States Olympic Committee (USOC) has identified overuse injury in elite athletes as a medical problem and the possibility of nutrient shortage as a part of the etiology. In May 1994, in a consensus decision, the USOC's Sports Medicine Committee stated that if "the athlete's dietary habits indicate possible reason for concern, prophylactic supplementation may be desirable." [Grandjean 1994, p. 1].

In Sweden, there is much academic debate concerning the average person's diet as well as that of risk groups for cardiovascular disease. Elite athletes have been added to the debate because the extent to which their diets are nutritionally sufficient has been questioned [Karlsson, Rasmussen, et al. 1992; Karlsson, Diamant, Edlund, et al. 1992].

In November 1993 the Swedish Nutrition Foundation (SNF), a non-government society, held a national hearing on the issue of whether vitamins and antioxidants are necessary. Participating at the hearing were all the major Swedish researchers in nutrition, representatives of all the Swedish government agencies involved in nutrition, representatives of all the major Swedish scientific bodies including The Royal Academy of Science, members of the pharmaceutical industry, and individual scientists engaged in related research programs. As a result of this hearing, a consensus decision suggested that risk groups for different disorders (e.g., cardiovascular diseases) could benefit medically from a megadosage of such nutrients as vitamin Q, vitamin E, vitamin C, and so on [Asp, Bruce, and Hambreus 1994]. In a recent consensus decision (December 1995), a Swedish sports administrative body corresponding to USOC's Sports Medicine Committee stated that (my translation) "If, on the other hand, an extreme energy need . . . is covered by carbohydrate-rich sources . . . the risk will increase for an insufficient supply of nutrients . . . . In such situations food supplement with minerals, vitamins, and fatty acids can also be of concern" [CPU 1995, p. 1]. This means that after 10 years of criticism and debate on my part, the Swedish sports medicine authorities have recognized the existence of the carbohydrate trap or fat_phobia, as Dr. Ann Grandjean, USOC, calls it. The carbohydrate trap represents a stage of malnutrition imposed by unprofessional advisors.

Inflammatory Processes and Nutratherapy

Although the biochemical evidence is convincing that radical biology is involved in inflammatory joint disease, myopathies, and overuse injury, very few nutraceutical therapy studies have been undertaken with a placebo-controlled design. Clinical support for nutratherapy is scarce, according to some researchers [Jenkins & Goldfarb 1993], whereas others believe that the clinical effects of such programs are evident [Kanter 1994].

It was recently reported that in a placebo-controlled intervention study of patients with rheumatoid arthritis, EFA, or vitamin F, improved the conditions of the patients [Leventhal, Boyce, and Zurier 1993]. The investigators administered the omega-6 fatty acid gamma-linolenic acid (GLA) at 1.4 g per day (see table 8.2). Cottonseed oil was applied as the placebo substance. Clinical endpoints were joint tenderness and swelling, morning stiffness, grip strength, and so on. No plasma EFA values, however, were obtained.

Vitamin F treatment of inflammatory joint diseases has resulted in clinical improvement, and plasma levels of the omega-3 and -6 series were determined [Drevon 1992; Horrobin 1989]. A suppression of the omega-6 fatty acid arachidonic acid (see table 8.2) has been suggested as a possible mechanism to reduce inflammatory symptoms. This might indicate that the origin of the disorder was not treated but rather that the symptoms were.

In a recent report, Keul and associates presented a nutratherapy study on vitamin E and elite racing cyclists. The protocol of a double-blind, placebo-controlled design was used. Although the nutratherapy did not improve performance, fewer signs of inflammatory processes were expressed as lower plasma muscle enzyme (CK, LDL, and GOT) activities after 5 months' nutratherapy [Rokitzki et al. 1994]. The authors, however, advised readers to be cautious with regard to conclusions about cause-and-effect relationships.

Placebo-controlled nutraceutical studies were also carried out concerning vitamin Q and its effect on physical performance at the Karolinska Institute in Stockholm, Sweden, in the early and mid-1980s. These studies were part of the preparation for clinical trials on cardiac failure and the possible effects of an adjuvant vitamin Q therapy. For legal reasons (secrecy agreements), the reports have not been published yet, but exercise data have been cited [Karlsson, Diamant, Folkers, et al. 1991]. Both muscle and plasma vitamin Q increased significantly. In addition, for those taking vitamin Q, exercise performance expressed as OBLA (from onset of blood lactate accumulation), exercise intensity, and symptom-limited performance were increased [Karlsson 1986a] compared to those taking the placebo.

In Scandinavia, in the mid- and late 1980s, food supplement programs were introduced to XC skiers as a prophylactic means. This approach was based on laboratory studies including placebo-controlled vitamin Q studies [Karlsson 1986b; Karlsson 1987]. At the start of the study, plasma values for vitamins Q and E were found to be low to very low in the XC skiers as compared to healthy sedentary controls (see figure 9.3a). Such clinical intervention programs to compensate for low plasma levels are referred to as *substitution therapy.*

Until the 1993-94 season, nutratherapy programs were not supported by the medical staff of the Swedish Ski Association, but they have gradually been accepted and applied by individual athletes. Plasma vitamin Q and E data on the national team level obtained from 1992 to 1994 showed marked changes and above-normal values as compared to the corresponding data from the mid-1980s (see figure 9.3b). Therapeutic levels for plasma vitamin Q are considered to be present in the range of 1.5 to 2.0 mg per L and above. The corresponding vitamin E levels are 17 to 25 mg per L and above [Karlsson, Diamant, Theorell, Johansen, et al. 1993].

When nutratherapy was suggested to the XC skiers in the mid-1980s as a prophylactic means, knowledge about vitamin Q, at least in the average medical community, was scarce. The extent to which vitamin Q therapy as an intervention violated the existing doping rules was therefore disputed. The same hesitation was not present with regard to the already well-known antioxidant vitamin E. To further document the plasma effects of vitamins Q and E, elite athletes (representatives of the Swedish national XC team) were treated with vitamin E only, while other elite XC skiers (who did not qualify for the national team) were treated with vitamin Q, under the surveillance of the Ethical Committee at the Karolinska Institute in Stockholm (see figure 9.4a and b).

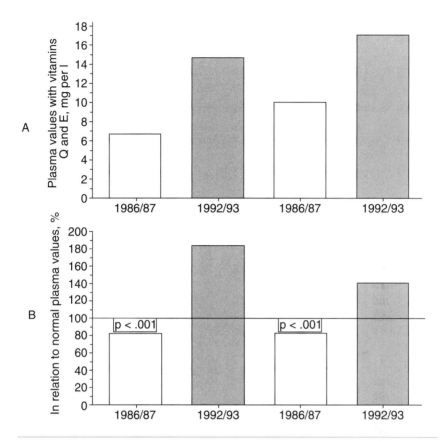

Figure 9.3   (a and b) Elite cross-country (XC) skiers from the Swedish national team. The absolute (a) and relative (b) plasma values for vitamin Q (on the left, multiplied by 10 in a) and vitamin E (on the right), before nutratherapy with antioxidant vitamins became accepted among athletes (1986-87), are compared to values of individual skiers after they had "unofficially" adopted similar programs (1992-93). The relative values represent percent values of healthy sedentary volunteers [Karlsson, Diamant, Theorell, Johansen, et al. 1993].

The vitamin E food supplement program disclosed marked elevations of plasma vitamin E values, whereas plasma vitamin Q remained below normal values (see figure 9.4a). The treatment with vitamin Q, however, resulted in elevated plasma values of both vitamins Q and E (see figure 9.4b) [Karlsson 1993].

The relationship between vitamin Q and vitamin E has been presented earlier (see figure 4.3). Vitamin Q as an endogenous entity evidently has the ability to protect vitamin E, whereas the opposite does not seem to be true. The catalyzing role of vitamin Q or the prosthetic-group-like property is evidently significant in many respects for vitamin E and the antioxidant defense system in general.

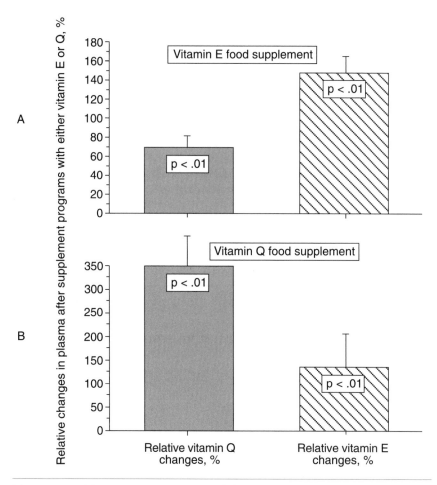

Figure 9.4   (a) For representatives of the Swedish XC ski team, results of their nutratherapy with vitamin E (200-300 mg a day) on plasma vitamin Q and E, expressed as the corresponding fraction of plasma values in healthy sedentary persons [Karlsson, Diamant, Theorell, Johansen, et al. 1993]. For reasons related to doping suspicions, vitamin Q was not administered. (b) Swedish XC skiers who were close to the elite level were administered vitamin Q (100 mg a day) for 6 weeks in accordance with a decision by the Ethical Committee at the Karolinska Institute, Stockholm, Sweden. The changed plasma values for vitamins Q and E are expressed as a fraction of the corresponding means in healthy, moderately active persons. [Karlsson, Diamant, Theorell, Johansen, et al. 1993].

Diet Antioxidants or Nutratherapy

The laboratory nutratherapy experiments in athletes were based on vitamins Q and E dissolved in small oil-containing capsules. Journalists used terms such as the "pill athletes" or the "pill sports" to indicate their attitudes. The results of the experiments sparked the question of whether a well-designed and balanced diet could provide, if not in the same amounts, at least a more efficient vitamin Q absorption and allocation. The rationale for such speculation was that these nutrients are present in and combined with their natural transporting moieties such as endogenous proteins and lipids in their food sources. This combination of these nutrients and their normal allies in the cell environment might represent a better bioavailability and subsequent absorption in the digestive systems.

An experimental diet was then designed with abundant carbohydrates and vitamin Q. After preliminary tests concerning preparation procedures, taste, and acceptability, it was applied during a week-long training camp among downhill skiers [Karlsson, Rasmussen, et al. 1992]. Twelve skiers of the Swedish national team, including one bronze-medal winner from the 1988 Winter Olympic Games in Calgary, were divided into two groups: a test group (TG) with the test diet, and a control group (CG) with a standard diet. Muscle biopsies and blood samples were repeatedly saved during that week and analyzed for muscle fiber types, muscle glycogen, muscle vitamin Q, plasma carbohydrates, lipids, and vitamin Q.

The two groups were of equal height, weight, muscle and vitamin Q content, and other features prior to the different dietary regimens. But after three days of heavy downhill ski training, muscle and plasma vitamin Q were suppressed in the CG (see figure 9.5a and b). The TG was expected to show elevated muscle and plasma vitamin Q values, as laboratory experiments had done with nutratherapy in combination with training. During and after the special diet, plasma and muscle samples were almost identical to data collected before training ("run-in data") (see figure 9.5a). It was concluded that the exclusive diet that the TG was served was disadvantageous with regard to absorption and allocation of vitamin Q to plasma and the contracting muscles, compared to the previous nutratherapy-supplemented diet based on antioxidant vitamins prepared in oil capsules.

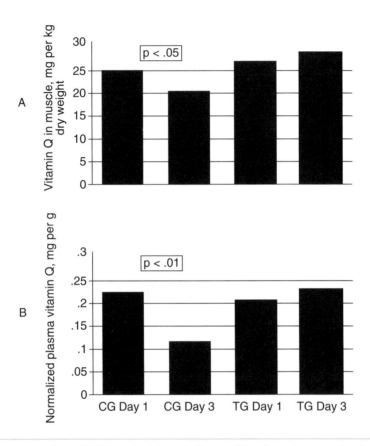

Comparative Studies on the Swedish
Downhill Skiing Team

Figure 9.5   (a and b) Muscle (a) and plasma (b) values for vitamin Q in the Swedish national downhill ski team during a 1-week-long training camp. The skiers were divided into two groups, one of which (test group, TG) was administered a special diet rich in vitamin Q and carbohydrates, and the other (control group, CG), given a normal diet [Karlsson, Rasmusson, von Schevelow, et al. 1992].

Antioxidant and Omega-3 Fatty Acid Nutratherapy

One of the major goals of the antioxidants' defense is to protect the PUFA, in general, and EFA, in particular. The two groups of EFA—the omega-3 (vitamin F_1) and omega-6 fatty acids (vitamin F_2)—are both active as membrane fluidizers and as precursors of different, and for each EFA, a specific eicosanoid synthesis (see figure 9.6a).

Of some concern is the omega-6 fatty acid arachidonic acid, which forms the more aggressive prostanoid 2-series (see figure 9.6b). Eicosanoids from the omega-3 series are generally less inflammatory than the omega-6 series. Arachidonic acid is synthesized within the omega-6 series (see table 8.2). The different omega-3 and -6 fatty acids and their syntheses depend on a number of shared enzyme systems, including desaturases and elongases. As precursors for eicosanoid synthesis, they are substrates for another set of enzymes—the cyclo-oxygenase and lipoxygenase enzymes [Mead and Mertin 1988]. This means that competition exists in a number of cases and that the availability of each precursor or substrate determines the outcome.

Arachidonic acid is abundant in the western diet. Moreover, the omega-3 fatty acids have been shown to be more susceptible to peroxidation than omega-6 fatty acids [Halliwell 1989]. As a result of the omega-6 fatty acid advantages, the ratio of omega-6 to omega-3 fatty acids (the Ω-6/Ω-3 ratio) in human plasma is 10 or more as compared to a desirable value of less than 5 [Skjervold 1991]. Clinically, there is another tool to describe the relationship between the EFA series: (EPA) x (AA)$^{-1}$ x 100 or (EPA/AA x 100) or the "relative EPA" (see table 8.2).

Because the omega-3 fatty acids are more susceptible to peroxidation than are the omega-6 fatty acids, they are more dependent on the antioxidant defense. This has been shown in elite XC skiers [Karlsson 1995]. Their Ω-6/Ω-3 ratio decreases on an individual level with plasma levels of vitamin Q (i.e., the more omega-3 fatty acids, the higher the plasma vitamin Q level). Antioxidant nutratherapy augments this effect along the same function (see figure 9.7a).

The relative EPA is negatively related to the Ω-6/Ω-3 ratio (see figure 9.7b). Therefore, omega-3 fatty acid EPA is of special interest in the suppression of omega-6 fatty acids, in general, and arachidonic acid, in particular.

EPA is especially rich in fatty, ocean fish species living close to the polar regions [Skjervold 1991]. Fish oil concentrate is manufactured from such sources. Fish oil concentrate is a nutraceutical product, commercially available since the 1980s. It is frequently confused with fish liver oil concentrate, a much older and more established nutraceutical.

The relationships between EPA intake, plasma EPA, and the relative EPA are presented in figure 9.8a and b. Obviously, a leveling off exists in the plasma EPA level with an increased EPA intake as a food supplement (see figure 9.8a). The saturation-like conditions have not been reported upon and the biological explanation is unknown. As a result, there is positively accelerating suppression of the omega-6 fatty acid arachidonic acid with an increased EPA intake (see figure 9.8b).

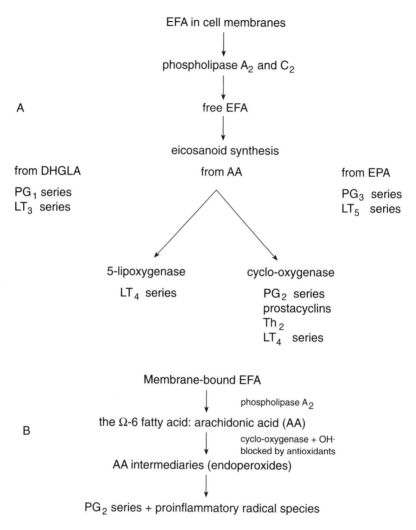

Figure 9.6 (a) Eicosanoid synthesis is dependent on EFA and the respective omega-3 and omega-6 fatty acids as precursors. The different prostanoid (prostaglandin), thromboxane, and leukotriene series are formed from them. EFA are released from their deposit site in phospholipid particles in the cell membranes. Certain enzymes, from which proteins are first formed when inflammation is present, catalyze the hydrolysis and the subsequent EFA release. (b) The arachidonic acid (AA) cascade is similar to the development of the inflammatory process. Radicals activate synthesis of phospholipase A_2 and the subsequent release of the omega-6 fatty acid AA. Antioxidants can block the cascade process at several locations.

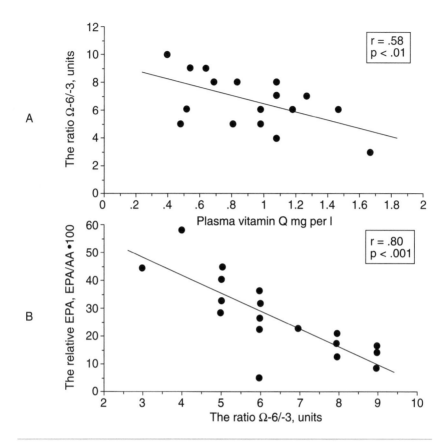

Figure 9.7 (a) On the x-axis, the plasma levels of vitamin Q before and after food supplements with antioxidants alone. On the y-axis, the corresponding individual relationships between plasma contents of omega-6 and the omega-3 fatty acids (the Ω-6/Ω-3 ratio). (b) The individual relationships between the Ω-6/Ω-3 ratio and the "relative EPA": [(EPA) x (AA)⁻¹ x 100 or EPA/AA x 100], for the same samples as presented in (a) [Karlsson 1995].

Antioxidants and Other Nutrients

It is well recognized that antioxidants have a protective role with regard to EFA in general and omega-3 fatty acids in particular [Haglund et al. 1991]. Reports show protective effects of antioxidants even toward such nutrients as glutathione, beta-carotene, some of the B vitamins, and selenium (Se). The role of the biological and pathological explanation for these observations can be multifactorial and, consequently, indirect in nature.

In studies undertaken in athletes and other groups, it has been confirmed that the plasma selenium content increases following antioxidant

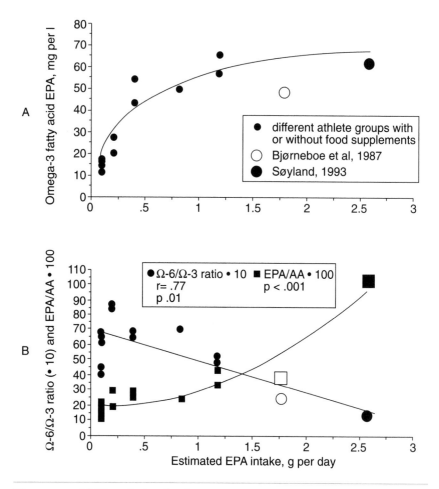

Figure 9.8   (a) The relationship between mean values for the estimated intake of the omega-3 fatty acid EPA in different studies and the corresponding plasma EPA mean [Karlsson 1995]. (b) The Ω-6/Ω-3 ratio and the "relative EPA" (EPA/ AA x 100) versus the estimated EPA intake as in (a).

nutratherapy (see figure 9.9) [Clausen 1991]. The increase can be quite substantial and amounted to 67% for the Swedish soccer team, which won the bronze medal in the World Cup in the summer of 1994 [Karlsson et al. 1994].

The blood samples of the Swedish soccer team were taken in April and May, before leaving for the World Cup in the United States. The second set of samples were taken when the team returned to Sweden in July of the same year. It is possible that the change in diet and food sources could at least partly explain the plasma selenium increase. Although the team brought with them their own cook and some food

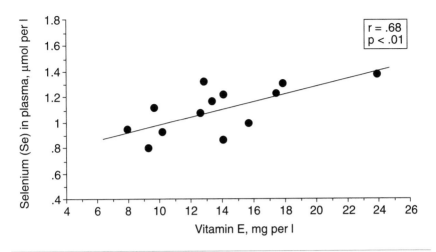

Figure 9.9   The relationship between plasma vitamin E means and the corresponding plasma selenium (Se) values [Karlsson 1995].

sources, most of what they ate was produced in the U.S. The North American soil is richer in selenium, which has a bearing not only on plants and their selenium content but also on livestock, for which these plants are a staple food source. The impact of selenium content in the ground on, for example, beef meat and, subsequently, on human health has been demonstrated by ground fertilization with selenium in Finland [Salonen, Salonen, and Seppänen 1988].

Summary

Short-time exercise (from less than 40 to 50 s) is limited by high lactates in muscle and blood, whereas performance tasks taking longer are limited by maximal pulmonary oxygen uptake ($\dot{V}O_2max$) or carbohydrates as the major fuel for heavy exercise. This means that net training (physically active) time, by definition, can be longer for the endurance athlete than for the sprinter. The impact of this circumstance on free oxygen-centered radicals ("spillover") from the mitochondria is obvious.

The energy output for physically active males engaging in leisure sport activities is 2,500 to 3,000 kcal a day and approximately 20% less for females. Male endurance athletes might have an average energy output on the order of 7,000 to 8,000 kcal a day. In relation to extremely competitive conditions (e.g., the 85 kilometer [53 miles] Swedish Vasa Cross-Country Ski Race, the Tour de France, and triathlons), the corresponding values might exceed 10,000 kcal a day.

The relative contribution of carbohydrates (CHO) expressed as energy equivalents (E%CHO) will increase curvilinearly, whereas the relative intake of protein (E%Pr) will remain unchanged (15% to 20%). As a consequence, lipid intake in relative terms (E%L) might decrease from the recommended 30% to 40% to values less than 25% with special dietary regimens. It is suggested that such dietary regimens carried out for long periods of time ("chronic treatments") are synonymous with malnutrition and an eating disorder and might be a threat to the athlete's health.

A diet rich in carbohydrates and poor in lipids endangers the availability of PUFA, in general, and EFA (vitamin F), in particular. The rationale for this suggestion is that the fat intake will (a) barely cover the fuel need, and (b) jeopardize the availability of structural fat to maintain normal biophysical and biochemical cell and membrane properties (PUFA in general) and precursors for eicosanoid synthesis (i.e., the EFA omega -3 and -6 fatty acids).

Cross-sectional studies on elite endurance athletes, carried out before nutratherapy programs became commercially available (i.e., before 1986-87), show marked plasma (lipoprotein) depletions of antioxidants such as vitamins Q and E. Moreover, their vitamin F plasma levels indicate an unfavorable ratio of omega-6 to omega-3 fatty acids.

The absolute plasma omega-6 fatty acid values are normal or elevated. As a consequence, the omega-3 fatty acid values are depressed. The omega-6 fatty acid arachidonic acid, especially, is readily available as a result of the livestock-meat-enriched diet in the urban western hemisphere. Arachidonic acid is the precursor of the inflammation-prone eicosanoids, especially the prostanoid (prostaglandin) 2-series.

Nutratherapy programs with vitamins Q and E increase the corresponding values both in skeletal muscle and plasma. Plasma values for the omega-3 fatty acids improve relatively more than omega-6 fatty acids. The ratio EPA/AA x 100, especially, has been found to be clinically relevant to demonstrate the favorable suppressing effects of arachidonic acid and the arachidonic acid cascade.

A combination of vitamin Q, vitamin E, and fish oil concentrate, which is rich in omega-3 fatty acids, in general, and EPA, in particular, has been demonstrated to optimize the desired plasma biochemical changes.

Vitamin Q and E food supplements have beneficial effects on other sections of the antioxidant defense mechanism, such as the selenium-glutathione and beta-carotene systems. There is no explanation for this, but it may be due either to improved absorption in the digestive systems or less pro-oxidative or peroxidative stress in different cells, tissues, and organs, causing less depletion and taxation of these systems.

10

Lipoidic Structures, Lipophilic Antioxidants, and Clinical Interpretations

As stated in chapter 3, most radical formation occurs within lipoidic structures, where the energy-releasing enzyme systems are assembled. It was also emphasized that even the first antioxidant-based defense line against reactive radical species was allocated to these lipoidic structures. The electrons/radicals were then transferred to the water phase and to the hydrophilic antioxidant systems, which constituted a later defense line (see chapter 4). This process was referred to as the antioxidant vitamin Q-E-C cycle. All these cycle entities were maintained in a reduced stage by the electron-feeding mitochondria. In this regard, vitamin Q has a key position in the electron translocation process.

Antioxidant Allocation and Its Significance

In previous chapters, vitamin Q and E contents have been expressed per gram muscle or liter of blood or plasma. It is obvious that this approach does not warrant a correct data interpretation. These nutrients as cell constituents are exclusively dissolved in lipoidic structures of the tissues. Moreover, the lipid content per gram muscle or per liter of blood will vary due to training status, dietary conditions, diseases, and so on.

In addition to allocation to lipoidic structures, transport proteins carrying these lipophilic compounds should also be considered as potential allocation sites. Based on experimental data, King and associates suggested a vitamin Q-specific transport protein for such purposes [King 1990; King et al. 1986].

Transport proteins are important in blood, for example, to offer water solubility to otherwise insoluble (hydrophobic) compounds. It is well established that the lipophilic vitamin A in blood is bound to transport proteins. If whole blood is compared to the corresponding serum or plasma fractions (i.e., a protein-free serum sample, as was seen in figure 6.6) complete identities exist between the vitamin Q contents (see figure 10.1). This means that all vitamin Q is allocated to the lipid phase of the plasma section of the blood—the lipoproteins, including the chylomicrons (see table 10.1 and figure 10.2).

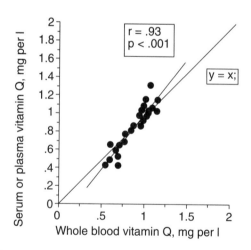

Figure 10.1   Vitamin Q in serum or plasma versus the corresponding (whole) blood sample. The illustration is a "line of identity figure." If no difference exists between serum or plasma data, on the one hand, and blood data on the other, all dots will gather on "line of identity," or y = x [Karlsson, Diamant, Theorell, and Folkers 1993].

Table 10.1   Plasma Lipids and Their Classification

Chemical structure	Compound	Presence
Lipoidic compounds		
Cholesterol	Cholesterol esters	Plasma lipoprotein particles (HDL, LDL, VLDL)
Ubiquinone		Lipoproteins + lipoidic membranes
Dolichol		Lipoproteins + lipoidic membranes
Tocopherols		Lipoproteins + lipoidic membranes
Fatty acids		
Stearic acid	Triglycerides (TG)	Free fatty acids (FFA) + TG
Oleic acid	TG	FFA + TG
EPA	TG	FFA + TG
DGLA	TG	FFA + TG
Esters		
Cholesterol esters		Lipoproteins (see above)
Di- (DT) or triglycerides	Phospholipids	Chylomicrons
(DG or TG, "true fat")		Lipoproteins + membranes
Phospholipids		Lipoproteins + membranes

Antioxidants in Different Organs and Tissues

An interorgan-intertissue comparison in healthy humans reveals that the higher the skeletal muscle vitamin Q content, the lower the blood vitamin Q content in a well-balanced, mixed diet (see figure 10.3a). Provided that vitamin Q is equally dispersed in the heart, skeletal muscle, and plasma lipid phases, a concentration gradient exists from the heart via skeletal muscle to the blood or plasma per unit of lipids. The highest content is present in the heart and the lowest in the blood (see figure 10.3b). This suggests an active or semi-active process involved in vitamin Q allocation to carry vitamin Q against the concentration gradient.

As expected, vitamin Q deposition to tissues and organs is also dependent on the level of metabolic activity in addition to the lipid content (see

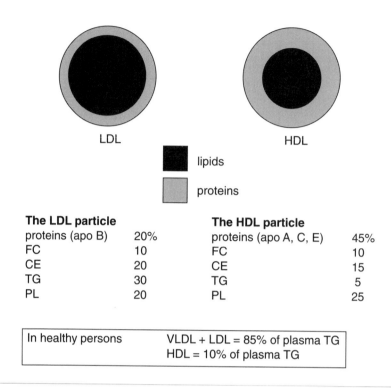

Figure 10.2   Schematic description of the different lipoprotein fractions in plasma and their contents in healthy persons. Very-low-density and low-density lipoproteins (VLDL and LDL) are similar in terms of contents and metabolic pathways. They are represented on the left by LDL, whereas the heavy-density lipoprotein (HDL) is depicted on the right. Their respective relative contents are expressed as a percentage of free cholesterol (FC), cholesterol esters (CE), triglycerides (TG), and phospholipids (PL). VLDL+LDL carry most of the total plasma TG.

figure 10.3c). Thus, per unit of weight, the liver and the heart are relatively richer in vitamin Q than, for example, respiratory and skeletal muscle. Adipose tissue is, of course, rich in fat, but from a metabolic point of view, it is relatively inert and, therefore, poor in antioxidants [Fredholm and Karlsson 1970].

The decrease in blood vitamin Q content with increasing muscle values is explained by the individual physical conditioning status [Nikkila, Kuusi, and Taskinen 1982; Stubbe, Gustafsson, and Nilsson-Ehle 1982]. The higher the training status, the higher the oxidative potential of the muscle tissue [Kiessling et al. 1973; Örlander, Kiessling, Karlsson, et al. 1977; Örlander, Kiessling, Larsson, et al. 1977] and the more vitamin Q deposition both as a coenzyme (CoQ_{10}) and as an unspecific antioxidant. Muscle lipids also increase the higher the oxidative or training status [Lithell et al. 1979; Karlsson 1979]. Blood lipids, on the other hand, decrease with physical training status [Nikkila, Kuusi, and Taskinen 1982], which means less volume to dissolve lipophilic compounds such as vitamins Q and E.

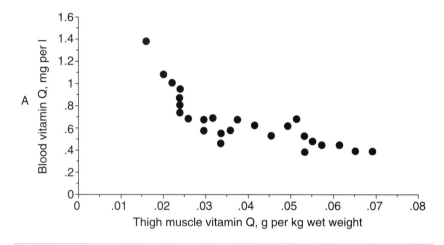

Figure 10.3   (a) The individual relationship between (whole) blood vitamin Q content and the corresponding vitamin Q value in the lateral portion of the thigh muscles (*m vastus lateralis*) in healthy persons representing different physical conditioning levels [Karlsson 1987].

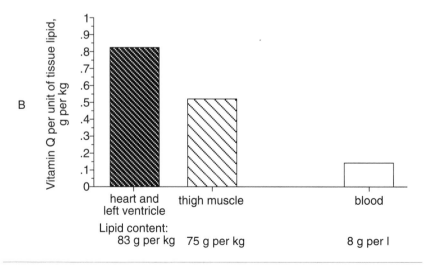

Figure 10.3   (b) Vitamin Q contents in heart muscle (left ventricle wall), skeletal muscle, and blood recalculated per unit of tissue lipid content in healthy persons [Lin et al. 1988; Karlsson 1987].

Tissue Vitamin Q Content in Healthy Persons

	Content	Volume or weight	mg	Amounts μmol	%
Plasma	≈1 mg/l	4l	4	4.6	0
Extracellular fluids	≈1	11	11	13	1
Heart	70 mg/kg	.4kg	28	33	2
Respiratory muscle	30	2	60	70	4
Skeletal muscle	40	25	1,000	1,159	71
Liver	60	2	120	139	9
Adipose tissue	10	20	200	232	13
Totals		65kg	1.4g	1.7μmol	100

Figure 10.3   (c) Approximate values for vitamin Q content in different organs and tissues. The approximate volume or weight of each organ or tissue is also presented. Based on these figures, the vitamin Q amounts have been computed and expressed in mg, μ-moles, and percent of total body vitamin Q content [Karlsson et al. 1993].

Physical conditioning status is partly related to both endowment and adaption. Individual variation in muscle fiber composition [Komi and Karlsson 1979] and adaptation to increased physical exercise [Saltin et al. 1968] are also interrelated. The higher the proportion of slow-twitch muscle fibers (%ST), the higher the increase in exercise capacity with training [Örlander, Kiessling, Karlsson, et al. 1977].

Muscle vitamin Q is consequently increased with an increased proportion of slow-twitch (ST), oxidative muscle fibers (%ST) (see figure 10.4a). The corresponding plasma vitamin Q values are, however, decreased for a group of healthy controls after a mixed diet because of the lowered plasma lipid content (see figure 10.4b) [Karlsson, Diamant, Theorell, Johansen, et al. 1993].

The same observations that have been made with regard to vitamin Q are also relevant to vitamin E [Karlsson, unpublished results].

Plasma Lipophilic Antioxidants and Their "Normalization"

The fact that plasma lipids vary on an individual basis should be taken into consideration when interpreting blood vitamin Q and E determinations. The vitamin Q and E contents should be related to the corresponding blood or plasma lipid content or a marker of that entity. This procedure (normalization) is a prerequisite for interindividual comparisons.

Plasma free cholesterol (FC) is a stable constituent in the different plasma lipoprotein fractions (see table 10.1) and, therefore, has been suggested as a marker for the plasma or serum lipid contents [Edlund 1988; Karlsson, Diamant, Edlund, et al. 1992; Johansen et al. 1991]. Plasma lipophilic constituents as vitamin Q are then related to this marker of the deposit volume.

If that is done for the group of healthy controls with a mixed diet, there is still a decrease in plasma vitamin Q versus skeletal muscle %ST (see figure 10.4c). This indicates that individuals with a high percent distribution of ST muscle fibers (%ST) have a relatively lower plasma vitamin Q saturation than individuals with a low %ST (i.e., individuals

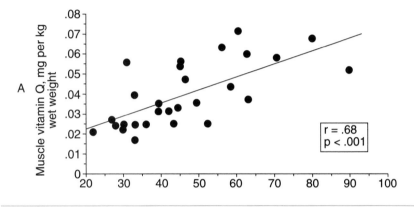

Figure 10.4a   Individual muscle vitamin Q content in relation to muscle fiber composition expressed as the relative distribution of slow-twitch (ST) muscle fibers (%ST) in the lateral portion of the thigh muscle (*m vastus lateralis*) [Karlsson 1987].

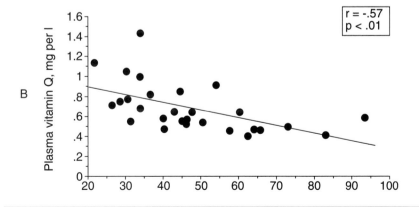

Figure 10.4b   Plasma vitamin Q content in relation to muscle fiber composition expressed as %ST for the same individuals as above.

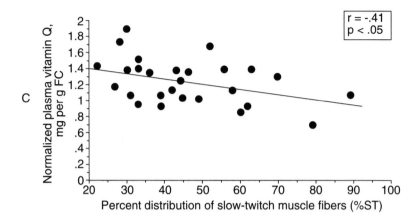

Figure 10.4c   Plasma vitamin Q content corrected for plasma lipid content ("normalized") for the same individuals as above in relation to muscle fiber composition, expressed as %ST [Karlsson, Diamant, et al. 1992]. Normalization has been performed based on simultaneous determination of plasma free cholesterol (FC) and ratio calculation: vitamin Q in plasma over plasma FC.

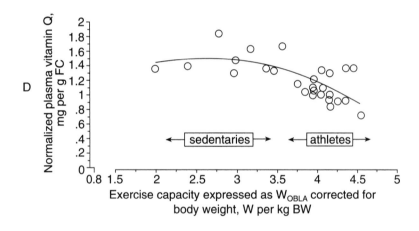

Figure 10.4d   Normalized plasma vitamin Q in relation to individual exercise capacity.

relatively rich in the other main fiber type, the glycolytic and fast-twitch (FT) muscle fiber) [Karlsson, Diamant, Theorell, Johansen, et al. 1993].

If the normalized plasma vitamin Q value after a mixed diet is related to exercise performance capacity, there is also a decrease in plasma vitamin Q and exercise capacity (see figure 10.4d). Sedentary individuals have high plasma vitamin Q values (around 1.5 mg vitamin Q per g FC),

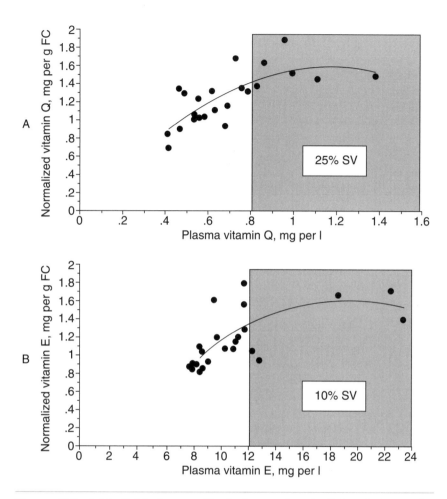

Figure 10.5 (a) Individual, normalized plasma vitamin Q in relation to the corresponding plasma vitamin Q value in healthy persons. (b) Individual, normalized plasma vitamin E in relation to the corresponding plasma vitamin E value in healthy persons [Karlsson, Diamant, Theorell, and Folkers 1993]. SV = saturated values.

whereas those involved in heavy endurance training might reach only 50% of that value. Hence, individuals with a higher exercise performance capacity (those rich in ST muscle fibers), have a plasma antioxidant reduction [Karlsson, Diamant, Theorell, Johansen, et al. 1993].

If normalized plasma vitamin Q and E values are related to the corresponding plasma values for individuals with and without nutratherapy programs with these antioxidants, there are negative curvilinear increases (see figure 10.5a and b) [Karlsson, Diamant, Theorell, Johansen, et al. 1993]. Those who have applied nutratherapy have plasma values 50%

to 100% higher than those who have eaten a mixed diet (mixed diet = 0.8 and 12 mg per L for vitamins Q and E [Karlsson, Diamant, Theorell, Johansen, et al. 1993]). It seems reasonable to assume, then, that the leveling off is a function of saturation-like conditions. Peak and saturated normalized values are, consequently, on the order of 1.5 to 2 and 25 to 30 mg per g FC or "units" for vitamins Q and E, respectively.

Thus, physically inactive individuals have saturated plasma vitamin Q levels, as depicted in figure 10.4d, whereas endurance trained athletes, for example, might be in a stage of relative depletion or exhaustion of these antioxidants. Therefore, those who are physically active have unsaturated plasma levels. The explanation for this could be that they have

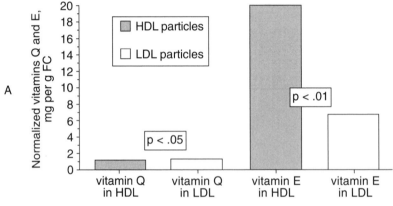

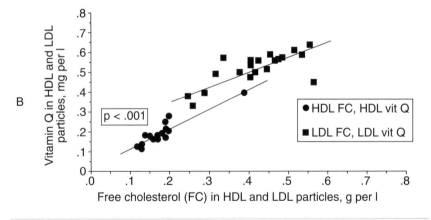

Figure 10.6   (a) Mean values for normalized vitamins Q and E in plasma HDL and LDL particles of healthy persons. (b) Individual data for vitamin Q in plasma HDL and LDL particle fractions versus the corresponding amount of free cholesterol (FC). The significance level denotes the differences on the y-axis for x = 0.

such a high turnover and breakdown of vitamins Q and E that a normal, well-balanced, and mixed diet cannot saturate the corresponding deposition volumes [Karlsson, Diamant, Theorell, Johansen, et al. 1993].

Some investigators have suggested that plasma lipophilic antioxidants can be related to the lipid volume or a corresponding marker for the purpose just presented [Horwitt et al. 1972; Okamoto et al. 1989]. Others suggest that these lipoidic structures could be involved in vitamin Q and E translocation from the viscera via the liver to different organs and tissue that benefit from or depend on these nutrients [Traber, Cohn, and Muller 1993; Traber and Kayden 1984; Karlsson, Diamant, Theorell, Johansen, et al. 1993]. These plasma lipids and their vitamin Q and E contents could also function as a source for reallocation of these antioxidants in a local metabolic or circulatory stress situation like myocardial infarction. This sort of reallocation has earlier been experimentally demonstrated in humans for glucose exchange mechanisms, where glycogen-exhausted muscles are re-fed from and by "rested" muscles and their normal glycogen content [Ahlborg and Jensen-Urstad 1991].

Vitamins Q and E in Plasma HDL and LDL Particles

Vitamin Q and E allocations are different in the two major plasma lipoprotein particles: the HDL and LDL (in this context, the sum of LDL and VLDL) particles. Whereas vitamin Q is more plentiful in the LDL than in the HDL particle per unit of FC, vitamin E is more abundant in the HDL particle fraction (see figure 10.6a). This distribution pattern is in line with the data of Traber and associates [Traber, Cohn, and Muller 1993].

If these lipoprotein particles are analyzed separately, the vitamin Q and E depositions depend on the lipid volume as depicted by the FC contents of the HDL and LDL lipoprotein fractions, respectively (see figure 10.6b).

The extent to which the higher vitamin E content in HDL bears on the clinically well-recognized cardiovascular benefits and less atheroma with plasma HDL or the ratio of HDL to LDL can only be speculated on. HDL is known as the "good cholesterol" or the "vacuum cleaner cholesterol." During the 1970s and 1980s, empirical clinical evidence was accumulated that indicated that both these plasma entities—HDL and vitamin E—are inversely related to cardiovascular disease and death [Gey 1986; Gey 1993; Kannel 1988; Wilson et al. 1987].

The causal explanation to atheroma and corresponding cardiovascular diseases has been that macrophages (see figure 6.6) and their scavenger receptors recognize peroxidized lipid material in plasma-borne LDL

Table 10.2   Content of the Plasma LDL Particle in Healthy Humans*

Per mol LDL (MW ≈ 2.5 million)	
1. Total amount of fatty acid (FA) molecules	2400
polyunsaturated FA (PUFA)	1200
omega-3 (Ω-3) FA	-
omega-6 (Ω-6) FA	130
2. Antioxidants	
vitamin E	6
vitamin Q	0.6
beta-(B-)carotene	0.3
3. Per unit of vitamin Q	
FA	5000
PUFA	2000
vitamin E	10
beta-(B-)carotene	5
4. Per unit of vitamin E	
FA	500
PUFA	200
vitamin Q	0.1
beta-(B-) carotene	0.5

*The normal content in plasma is approximately 2 g per l.

particles. After consuming them, migrating, and penetrating into the endothelium, they undergo a histological change—the "foam cell" formation. This is an inflammatory process and the situation—the atheroma or the atherosclerotic lesion—is referred to as an *atherosclerotic process*. This process underlies a major theory as to the causal events of this devastating cardiovascular disease in the western hemisphere [Goldstein and Brown 1987; Brown and Goldstein 1983; Ravnskov 1992].

Vitamins Q and E
and Their Transport Vehicle—LDL

Whereas all cell systems investigated can form their own vitamin Q [Appelkvist, Kalén, and Dallner 1991], a transport system must be present for vitamin E between the digestive systems and the cells and tissues in need of this essential nutrient.

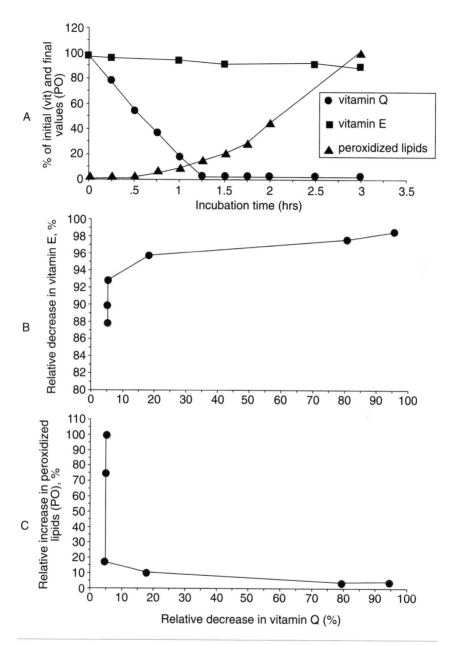

Figure 10.7   (a) A presentation of the data of Stocker and associates with respect to *in vitro*-provoked peroxidation of human LDL particles [Stocker and Frei 1991]. The values are expressed in percent of initial vitamin Q and E contents and final concentrations of peroxidized lipid material (PO). (b and c) The data in (a) are reorganized to illustrate the "all or nothing" type of relationship between the disappearance of vitamin E and the appearance of peroxidized lipid material (PO).

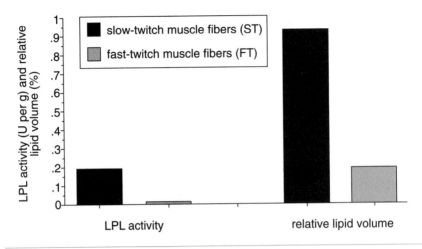

Figure 10.8   Lipoprotein lipase (LPL) activity and relative lipid droplet volume in human thigh muscle (*m vastus lateralis*) [Lithell et al. 1979; Lithell et al. 1982].

The suggested vitamin Q and E transport vehicle LDL [Karlsson, Diamant, Theorell, and Folkers 1993; Traber, Cohn, and Muller 1993] and its contents are presented in table 10.2. It is obvious that vitamins Q and E "control" several thousand fatty acids and especially PUFA.

The significance of the lipoprotein-borne antioxidants in protecting the LDL lipoprotein particle from lipid peroxidation has been repeatedly demonstrated by several groups [Frei et al. 1988; Stocker, Bowry, and Frei 1991; Stocker and Frei 1991; Bowry and Stocker 1993; Esterbauer et al. 1987; Esterbauer et al. 1990; Esterbauer et al. 1991]. Stocker and associates showed experimentally and *in vitro* that peroxidized material in the LDL particle appeared first, when both vitamins Q and E had been decreased (see figure 10.7a) [Stocker, Bowry, and Frei 1991]. If their data are reorganized, it is possible to show how both the disappearance of vitamin E (see figure 10.7b) and the appearance of peroxidized lipids (see figure 10.7c) follow "all or none response" relationships related to vitamin Q content.

The vitamin Q and E allocation ability of the LDL particle has a match, or common denominator, in the periphery as a means for transport of lipids or lipophilic compounds—the endothelium-bound lipoprotein lipase (LPL) receptor. The ST fiber-rich skeletal muscle has a higher muscle (endothelium-bound) lipoprotein lipase activity than the FT fiber-rich muscle (see figure 10.8) [Lithell et al. 1981; Lithell et al. 1979]. Skeletal muscle LPL enzyme is synthesized by the muscle fibers, released to the extracellular space, and then attached to the endothelium [Nilsson-Ehle, Garfinkel, and Scholtz 1980].

The LPL enzyme, or rather the endothelium-bound receptor containing LPL properties, hydrolyses LDL-borne triglycerides (lipolysis), which furnishes the muscle with the necessary fatty acids for fuel deposition as triglycerides. The necessary glycerol molecule for triacyl formation can either be obtained from the bloodstream or endogenously synthesized in the muscle from three carbon residues as lactate, for example [Fredholm and Karlsson 1970]. The net result of this is a higher fat content in ST muscle fiber than in FT muscle fiber despite higher activities for fat-degrading enzyme systems in the ST fiber [Lithell et al. 1979; Karlsson 1979].

Vitamins Q and E and Their Turnover: A Role for HDL?

Most cholesterol in humans is a product of an endogenous synthesis. It is equally well recognized that the HDL particle takes up cholesterol in the periphery and transports it to the liver, where it is either metabolized or released to the digestive system as bile. The metabolism and the release are two major determinants of the total cholesterol turnover in humans.

Why do humans, and all organisms, treat an endogenous metabolite in such a wasteful way? Moreover, does the body treat its endogenous ubiquinone in the same fashion? If the answer to the latter question is yes, it seems reasonable to suggest that vitamin Q and vitamin E are handled in the same way as cholesterol. As HDL takes up cholesterol in the periphery, it is possible that HDL also takes up vitamins Q and E for metabolism in the liver and a release to the digestive system.

One teleological explanation for this phenomenon could be that it grants a continuous risk of depletion in different organs and tissues. This could be a guarantee of a satisfactory turnover of biological material. The threat could serve as constant stimuli to maintain an appropriate endogenous production or uptake of these nutrients in the digestive systems.

Endowment and Training Adaptation and Antioxidants

The allocation and turnover mechanisms of vitamins Q and E to the periphery are the results of these characteristics:

- Endowment with respect to muscle quality and consequent performance characteristics

- Promotion of different physical performance traits, such as endurance activities
- Consequent adaptive responses both within and outside the cell
- An individually different oxygen turnover and need for antioxidants

These biological mechanisms will make up the different properties referred to as *physical conditioning*, or *training effects*. Sequential features related to the periphery and endurance training effects can be summarized in this process:

1. Increased physical conditioning level
2. Increased mitochondrial and respiratory activity
3. Increased capillary density and local molecular oxygen supply
4. Increased need for antioxidants
5. Increased muscle LP lipase synthesis
6. Increased deposition of fat (fuel) and antioxidants
7. Increased site allocation of muscle LP lipase
8. Increased endothelium volume related to capillarization
9. Increased capillary density due to training
10. Back to 1, and continue

It is of course debatable as to which came first, the chicken or the egg. It has been shown that endowment is a significant factor determining muscle fiber composition [Komi and Karlsson 1979]. Capillarization and, consequently, local blood flow and molecular oxygen supply increase with muscle fiber composition expressed as percent distribution of ST muscle fibers (%ST) [Andersen 1975; Thompson et al. 1988]. Ergoreceptor activity is also a local skeletal muscle-fiber-related feature, which improves regulation of central circulation and, indirectly, pulmonary activity the higher the %ST [Victor et al. 1988; Mitchell et al. 1977; Shepherd et al. 1981; Mitchell 1985]. As a result, the ST fiber-rich muscle has a more efficient regulation of both central and peripheral circulation regulation.

In endurance athletes, the FT muscle fiber is more susceptible to injury than the ST fiber (figure 2.1b and c) [Wallensten and Karlsson 1984a, b; Wallensten and Eriksson 1984]. This could be a basis for an active disqualifying factor in the recruitment process of individuals in endurance sport activities.

Thus, a critical feature in endurance athlete recruitment is the genetic code determining the individual muscle fiber composition expressed as %ST muscle fibers [Karlsson 1979] and the nutritional status providing the individual with ample amounts of antioxidants [Witt et al. 1992].

Summary

Vitamins Q and E in blood plasma decrease with skeletal muscle increases of the same antioxidants. This inverse relationship is largely explained by the fact that training increases muscle deposition but decreases the lipid deposit volume in plasma.

To make interindividual comparisons meaningful, plasma values of these lipophilic antioxidants should be related to their deposit site (i.e., the lipid volume has to be considered). This procedure to correct for the deposition volume is frequently referred to as normalization.

There are different means to normalize for the lipid volume. Free cholesterol has been chosen because it is a stable entity of plasma lipids and is present in all lipoprotein fractions.

It is concluded from comparative studies that saturated plasma levels for vitamins Q and E correspond to 1.5 to 2 and 25 to 30 mg per g of FC or "FC units," respectively.

There is a decrease in normalized plasma vitamins Q and E versus either percent distribution of ST muscle fibers (%ST) or exercise performance capacity after a period of time with a mixed, well-balanced diet. Plasma lipid saturation values as low as 50% have been documented. It is suggested that this represents a situation of relative depletion or exhaustion of these nutrients in persons rich in ST muscle fibers or those with a high exercise capacity. A high exercise performance capacity corresponds to a high daily energy turnover. Energy turnover as occurs in physical training implies a vitamin Q and E breakdown.

Deposition of vitamin Q is more apparent in LDL, whereas vitamin E dominates in HDL particles. This may be relevant to the positive effects of HDL in reducing cardiovascular diseases.

LDL particles have been thought to be transport vehicles of lipophilic antioxidants from the digestive system to organs and tissues and between organs and tissues.

Local muscle LPL receptor activity could act as a receptor for both fat and lipophilic antioxidant deposition in skeletal muscle to meet high physical-activity-related demands on both fuel and antioxidant supplies.

III

Nutratherapy and Sports Medicine

How does the knowledge gathered thus far apply to fitness and elite sport? Chapter 11 reports facts from well-documented nutratherapy studies in sport. The reader might seek references to studies that either disprove the effects of nutratherapy or show deleterious effects. Such reports have been published, but frankly, I have not found them to be academically meritorious. I have also excluded studies of poor quality that favor the concept of nutratherapy. Fitness and elite sports are equally dependent on good nutrition standards; nutratherapy is a means to achieve that.

As is true for pharmaceutical treatments, dose-response relationships and side effects have to be considered in nutratherapy. These concerns are covered in chapter 12. Recommended Daily Allowances (RDA)–supplement levels suggested by different national and international bodies–are discussed when available. Their relevance has frequently been questioned. The lipophilic vitamins Q and E increase in muscle and blood according to individual characteristics, probably related to individual bioavailability variables. The suggested goal of nutratherapy is achievement of saturation levels of plasma lipoproteins. General recommendations can be presented for recreational and semi-elite athletes for vitamins Q, E, and F_1. Elite athletes should monitor the progress of their nutratherapy through blood level monitoring. Decreased blood viscosity has been reported as a side effect of vitamins Q, E, and F_1. This enhances capillary blood flow which is advantageous in oxygen delivery, but may reduce blood clotting activity. Patients being treated with platelet-reducing drugs should consult their physicians before beginning nutratherapy. Another side effect is that antioxidants can, with overdosing, turn into prooxidants. This has been demonstrated for vitamins C and E, and is theoretically possible, but biochemically unproven, for vitamin Q.

Chapter 13 introduces the reader to the federal regulations that ensure safe nutratherapy. In many respects, nutraceutical therapy is similar to pharmaceutical therapy, and the consumer has the right to demand the same rules for safety and valid information. Legislation has empirically been found to be the only satisfactory way to grant the consumer a safe nutratherapy.

According to chapter 13, nutratherapy has nothing to do with doping as the term is used today. International societies, including those for sports, should act more responsibly to further nutratherapy as a means to improve the quality of life of average people as well as of fitness athletes. Nutratherapy also offers the elite athlete proper medical care, which will grant him or her health and optimal sport performance while in the public eye, but it will also offer the retired athlete a pleasant, healthy retirement.

CHAPTER 11

Relevant Studies

Based on biochemical and cell-physiological evidence, the clinical effects of nutratherapy with antioxidant vitamins should result in these qualities:

1. Less overuse injury and faster recovery from an inflammatory process
2. Improved immune response and faster recovery after an infection
3. Less disease in the fitness athlete and a more efficient training program in the elite athlete
4. A better quality of life for the fitness athlete and improved sports performance for the elite athlete

Antioxidant Vitamins and Placebo-Controlled Studies

Studies have been conducted claiming that nutratherapy with antioxidants improved physical performance directly or indirectly [Buzina & Suboticanec 1985; Dillard et al. 1978; Kanter, Nolte, and Holloszy 1993; Karlsson et al. 1994; Karlsson 1995; Karlsson, Diamant, Folkers, et al. 1991; Karlsson, Diamant, Theorell, et al. 1991; Packer and Viguie 1989; Simon-

Schnass and Korniszewski 1990; Simon-Schnass and Pabst 1988; Simon-Schnass, Reiman, and Böhlau 1984; Sumida et al. 1989; Witt et al. 1992].

Placebo-controlled studies have also shown that vitamin E nutratherapy improves the immune response to infectious diseases [Meydani, Barklund, and Liu 1989; Meydani, Hayek, and Coleman 1992].

The beneficial effects on the immune response have also been demonstrated for omega-3 fatty acid (fish-oil concentrate) nutratherapy [Søyland et al. 1993; Drevon 1992; Berg Schmidt et al. 1990; Berg Schmidt et al. 1991].

No studies have been published in which the placebo-controlled efficacy of vitamin Q was tested with regard to overuse injury or the immune response. It should be noted, however, that vitamin Q was discovered in the 1950s [Crane et al. 1957; Morton et al. 1957]. Nutraceutically, vitamin E is more than 50 years older than vitamin Q. Furthermore, the nutraceutical values of vitamin Q were only beginning to be clinically appreciated in the mid-1980s [Beyer and Ernster 1990; Karlsson et al. 1990b; Mortensen, Heidt, and Sehested 1990; Mortensen et al. 1985]. By that time, the placebo-controlled prospective nutratherapy studies with vitamin E in relation to cardiovascular diseases, for example, were already in the planning stages [Stampfer et al. 1992; Rimm et al. 1992].

Table 11.1 lists published placebo-controlled studies of antioxidant supplements where nutraceutical efficacy could be demonstrated on chosen clinical endpoints.

Placebo-controlled studies have revealed less lipid peroxidation with antioxidant supplement programs at both normal and high altitudes (mountaineering) [Simon-Schnass and Pabst 1988; Meydani et al. 1993]. Improved exercise performance capacity is also present according to two placebo-controlled studies (see table 11.1) [Simon-Schnass and Pabst 1988; Karlsson, Diamant, Folkers, et al. 1991].

In a recent report, Keul and his group investigated the long-term effect of vitamin E nutratherapy in elite German racing cyclists. They found signs of fewer inflammatory processes but no effect of the nutratherapy on performance capacity [Rokitzki et al. 1994].

Antioxidant Supplements and Open Studies

Clinical effects of nutratherapy with antioxidants in open studies, either with a matched control group or with the subjects as their own controls, corroborate the results of the previously mentioned studies in improved exercise capacity and less lipid peroxidation (see table 11.1). Less damage to genetic structures (in this case, RNA) has also been reported [Witt et al. 1992].

Table 11.1   Antioxidant Intervention Studies of Healthy Persons and/or Elite Athletes, Published in Scientific Journals or Available as Good Clinical Trial Practice Reports From the Pharmaceutical Industry

Author(s)	Type of intervention	Period	Clinical endpoint(s)
A. Placebo-controlled studies			
Simon-Schnass and Pabst 1988	Vitamin E, 400 mg a day	4 weeks	Exercise capacity at altitude, lipid peroxidation, pentane expiration
Karlsson, Diamant, Folkers, and Lund 1991	Vitamin Q, 100 mg a day	6 weeks	Exercise capacity
Kanter, Nolte, and Holloszy 1993	Vitamin E, 727 mg; C, 1,000 mg; β-car, 28 mg a day	6 weeks	Endurance exercise, lipid peroxidation, serum MDA, pentane expiration
B. Studies with matched controls			
Simon-Schnass, Reiman, and Böhlau 1984	Vitamin E, 400 mg a day	6 weeks	General deterioration
Packer and Viguie 1989	Vitamin E, 800 mg; C, 1 g; β-car, 10 mg	2 months	Muscle trauma, plasma CK increase
Sumida et al. 1989	Vitamin E, 300 mg a day	4 weeks	Maximal exercise, lipid peroxidation and cell trauma, blood MDA and enzymes
Simon-Schnass and Korniszewski 1990	Vitamin E, 400 mg a day	4 weeks	Altitude and blood viscocity (filterability) cell trauma, less protein loss

(continued)

Table 11.1   *(continued)*

Author(s)	Type of intervention	Period	Clinical endpoint(s)
Witt et al. 1992	Vitamin E, 500 mg; C, 1g; β-car, 10 mg	1 month	Endurance exercise, RNA damage, 8-HOG
C. Open studies			
Dillard et al. 1978	Vitamin E, 1,091 mg a day	2 weeks	Submaximal exercise, lipid peroxidation, pentane expiration
Pincemail, Deby, and Dethier 1987	Vitamin E, 200 mg a day	3 weeks	Endurance exercise, pentane expiration
Buzina and Suboticanec 1985	Vitamin C, 80–1,000 mg a day	Acute administration	Exercise capacity
D. Open GCTP studies in Sweden			
Karlsson et al. 1994	Vitamin Q, 100 mg; E, 300 mg	1.5 months	Infectious disease status, plasma antioxidants, and omega-3 and -6 fatty acids
Karlsson 1995	Vitamin Q, 100 mg; E, 300 mg	4 months	Infectious disease status, plasma antioxidants, and omega-3 and -6 fatty acids

Good Clinical Trial Practice Studies

The field of nutratherapy in medicine is quickly developing. The knowledge and the experience gathered in one field (for example, cardiology and adjuvant nutratherapy in cardiac failure), should be transferred to other fields as quickly as possible. The need for relevant prophylactic and rehabilitative measures for fitness and elite athletes is well documented, as is the eagerness of these athletes to apply and test recently developed knowledge. Sometimes this eagerness has led to accusations of applying unethical means (even doping) to artificially improve performance.

The extent to which this form of "science diffusion" needs a repetition of placebo-controlled studies has been discussed in clinical pharmacology. The alternative would be that a concept of a treatment can be taken for granted when it has once and for all been proven, and there is a need only of a "dose-response" study in a new application field. Such simplified studies in a new patient population will provide a quick basis for an optimal dosage program, administration frequency, and procedures. There is no general solution to this problem. I am, however, reluctant to accept the ethical consequences of postponing an adequate medical treatment for several years due to the lack of a field-related, double-blind, crossover study and its evaluation by the science community or government agencies.

I have initiated or participated in nutratherapy studies in elite sports, in collaboration with the pharmaceutical industry, on both sides of the Atlantic. As the basic scientific background for these studies has been well documented and accepted, I have taken the responsibility to restrict them to open clinical studies. That does not mean I oppose further academic studies; I would be happy if data and experience gathered from these open studies could be used to improve future blind and double-blind studies, with or without crossover profiles.

The goal of these nutratherapy programs has only been to (re-)establish normal blood-plasma values for some of the nutrients in question. Thus, from a formal point of view, they have been substitution therapies with antioxidant vitamins alone or in combination with vitamin F_1 (fish-oil concentrate rich in omega-3 fatty acids). Together with team physicians, I have followed normal clinical procedures to register symptoms and diseases in the medical records of the individual athletes.

Taken altogether, these studies have followed what in clinical pharmacology and therapeutic product development is referred to as *good clinical trial practice (GCTP)* [NLN 1993]. Some of these studies and the corresponding data have been released, and I report them next.

Nutraceutical Preparation of the 1994 Swedish World Cup Soccer Team

The Swedish national soccer team sought advice with respect to its food intake and supplement programs as a part of their preparations for the World Cup of 1994 (see figure 11.1). They had a notion that their recommended food intake was not optimal and that the corresponding food supplement program was inadequate. Team advisers and consultants had recommended a carbohydrate-enriched long-term diet, and the team was caught in the "carbohydrate trap" (see chapter 2), a frequent syndrome in current elite sport activities. The relative carbohydrate intake expressed as a percent of the total energy intake might on an individual level even exceed 60% (E%CHO), whereas the lipid intake could be less than 20% (E%L), at least occasionally.

A change in diet was initiated in May 1994 so that the lipid intake (E%L) was increased to 30% to 35%. Simultaneously, blood samples were saved for baseline ("run-in") plasma values of vitamins Q, E, and F (i.e., the omega-3 [F_1] and omega-6 [F_2] fatty acids). A general nutratherapy was then initiated with 100 mg of vitamin Q, 300 mg of vitamin E, and 2 g of vitamin F_1 per day. The program was based on pharmaceutical-grade products. For individuals with extremely low, or for other reasons deviating plasma values, special supplement programs were administered [Karlsson et al. 1994].

For 16 of the 24 players, run-in plasma vitamin Q was less than the mean for healthy persons (0.8 mg per liter) [Johansen et al. 1991; Karlsson, Rasmusson, et al. 1992; Karlsson, Diamant, Theorell, Johansen, et al. 1993]. Three players had more than 1.0 mg per liter, which indicated a previous supplement program including vitamin Q. The corresponding vitamin E values were even more depressing: All players had plasma values less than 12 mg per liter. Seven players had Q less than 0.6, and 14 players had E values less than 8 mg per liter.

The earlier documented linear relationship between vitamins E and Q was confirmed, but on a lower level than for healthy controls (see figure 11.1a). Seven players had vitamin Q and E relationships lower than 95% of the confidence intervals for the corresponding means (see figure 11.1b). The ratio of plasma vitamin E to Q was elevated (greater than 15) for those who were poor in plasma vitamin Q. The ratio then decreased (a negative curvilinear decrease) as plasma vitamin Q values increased (see figure 11.2a), which was documented earlier (see figure 11.2b). For those who were well nourished or supplemented, the ratio ranged from 5 to 10.

As described in chapter 3, a major antioxidant defense activity develops against polyunsaturated fatty acids, in general, and essential fatty acids, in particular (see table 8.2).

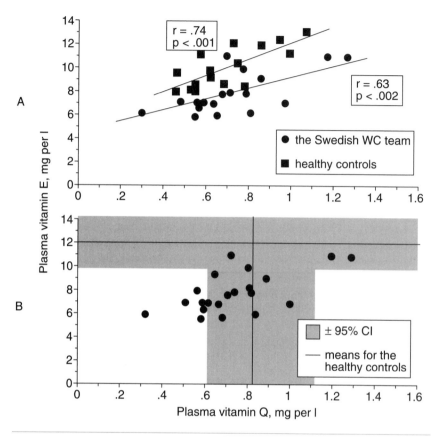

Figure 11.1 (a) The Swedish World Cup (WC) soccer team of 1994 and their individual plasma vitamin Q and E values before the start of nutratherapy ("run-in" data, May 1994). For comparison, corresponding data for healthy, moderately active controls are included [Karlsson, Branth, and Ekstrand 1994; Karlsson, Diamant, Theorell, et al. 1993]. (b) Mean values for plasma vitamins Q and E for the soccer players in (a) and the corresponding 95% confidence interval (CI) area.

Nutritionally speaking, the plasma omega-3 and omega-6 fatty acid values in the soccer players were equally as disturbing as the antioxidant data. Relative EPA (EPA/AA x 100, see table 8.2), for 16 players was less than the mean for healthy controls (30%) and far from the therapeutic level (>50%) (see figure 11.3). This is indicative of an elevated plasma omega-6 level, in general, and arachidonic acid (AA, see table 8.2), in particular. Arachidonic acid is the precursor of the more aggressive and proinflammatory prostaglandins (prostanoids) (see figure 9.6a).

After returning from the United States with the bronze medal (third place), the Swedish soccer team had a new set of blood samples taken.

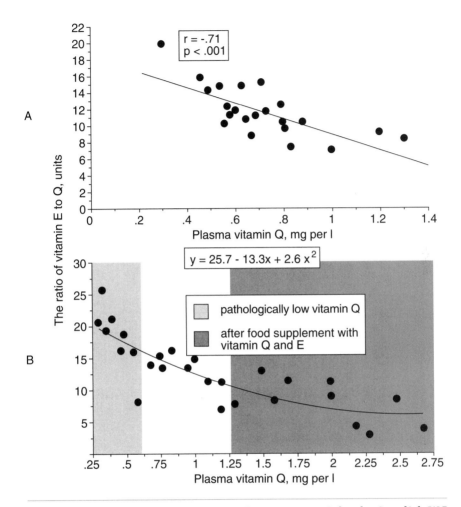

Figure 11.2 (a) The individual ratio of vitamin E to Q for the Swedish WC soccer players versus the corresponding vitamin Q values (see fig. 11.1a and b). The regression line denoted assumes a linear relationship between the ratio and vitamin Q. (b) The ratio of plasma vitamin E to Q for patients with arthritic-like muscle and joint pain before and after a nutraceutical intervention with vitamins Q and E for 6 months. The best-fitting curve is denoted.

Plasma vitamin Q and E had increased 26% and 39%, respectively, and relative EPA averaged 26%.

The following effects of the nutratherapy were observed. During the first 30 days with the original dietary regimen, only six players were free from any kind of disease symptom, and six players had infectious diseases (four of whom were treated with antibiotics). During the

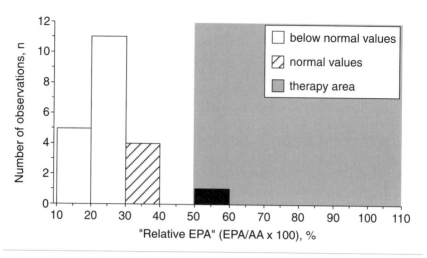

Figure 11.3 Individual observations of "relative EPA" (EPA/AA x 100) before the start of nutratherapy in the soccer players of the 1994 Swedish WC team. Only one player had a relative EPA corresponding to the therapy area (the solid black bar) [Karlsson, Branth, and Ekstrand 1994].

remaining 50 days with the new diet and an adequate nutratherapy, only one player had an infection that was treated with antibiotics (see figure 11.4a). In addition, less overuse injury was reported in the latter 50 days than in the previous 30 days.

In the month of May 1994, for example, there was a linear relationship between the reported number of days off due to disease or injury and the vitamin E over Q ratio (see figure 11.4b). This ratio describes the nutraceutical quality of the antioxidant defense system: The better the antioxidant defense, the lower the ratio. Consequently, the better the antioxidant mechanism, the fewer days missed from the training and competition schedules.

Cross-Country Skiing, Plasma Antioxidants, and Food Supplements

As described in chapter 2, the Swedish cross-country (XC) ski elite developed an addiction to carbohydrates during preseason training and the season itself, covering the months of October to April (see figure 2.4). The mean value for lipid intake during the 1992-93 preseason was 27% energy (E%L), which would cover only the most immediate needs of lipids as a fuel.

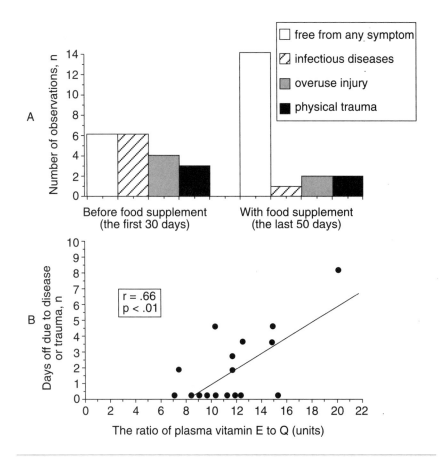

Figure 11.4   (a) Disease history of the Swedish WC soccer team during the preparation for and the tournament in the U.S. During the first 30 days, the players maintained their old dietary regimen. The diet was characterized by an extremely high carbohydrate (CHO) content and a low fat content. During the last 50 days, including the day of the semifinal game, the CHO content was reduced and the lipid intake correspondingly increased to an average of 30% to 35% energy (E%L). In addition, a nutratherapy based on vitamins Q, E and F_1 (omega-3 fatty acids) was introduced. Four of the six players who got infectious diseases during the first 30 days needed treatment with antibiotics, whereas the one player who got an infectious disease in the last 50 days needed such treatment. (b) The relationship between the number of training days or games missed due to infectious diseases or physical injury reported for May 1994, and the plasma antioxidant status expressed as the ratio of plasma vitamin E to Q (see figure 11.2b). These data were obtained before the nutratherapy was initiated.

In the mid-1980s, I introduced a food supplement with vitamins E and C at the megadosage level to elite Swedish XC skiers. This nutratherapy program was ethically accepted by the team physicians. Even if depressed plasma vitamin Q levels also were demonstrated, such a nutratherapy was banned as a doping intervention.

As a result of the XC skiers' dietary habits, most had plasma vitamin Q and E deficits (see figure 11.5a and b). For those who had no nutratherapy, the means were 0.6 and 9.6 mg per L, or 75% and 80% of the means for healthy controls [Karlsson, Diamant, Theorell, Johansen, et al. 1993]. None of those who had a nutratherapy program of their own

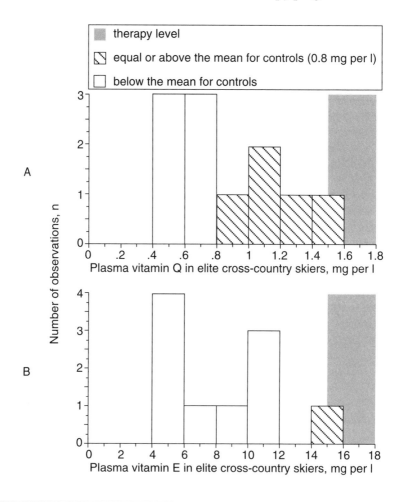

Figure 11.5   (a and b) Elite XC skiers and their individual plasma vitamin Q and E levels, presented as frequency illustrations [Karlsson 1995].

reached the therapeutic levels (see chapter 9). This failure to reach therapeutic vitamin Q and E levels could be explained by inadequate administration, but also could be due to poor quality of the health food store products, which has been frequently demonstrated [Cui et al. 1994].

Two intervention programs (Study I and Study II) have been tested on XC skiers, based on

a. vitamins Q and E (100 and 700 mg a day, respectively) and
b. vitamins Q, E, and F_1 (4 g per day of fish-oil concentrate were added to the vitamins administered according to [a]).

For both programs, satisfactory plasma vitamin Q and E levels were obtained after 2 to 3 months of therapy. Plasma vitamin E, however, was found to be lower in Study II than in Study I (13.3 and 17.5 mg per L, $p<0.01$). This is in line with previous findings where supplements with omega-3 fatty acids have been found to depress plasma vitamin E [Haglund et al. 1991; Bjørneboe et al. 1987]. Evidently, the addition of the peroxidation-prone PUFA to the nutratherapy program demanded more antioxidant support than the one actually administered (700 mg of vitamin E per day).

As seen in the Swedish World Cup soccer team, nutratherapy decreased the ratio of vitamin E over Q versus plasma vitamin Q (see figure 11.2) even in the XC skiers (see figure 11.6a and b). Before and after Study I, negative relationships were present, but the curves were shifted (see figure 11.6a). Assuming a polynomial curve, the best-fitting curve showed a negative relationship (see figure 11.6b) with a plateau corresponding to a vitamin E over Q ratio of 5 to 10 for a plasma vitamin Q content in the therapeutic range.

As a result of the antioxidant nutratherapy of Study I, it could be seen that antioxidants protected the PUFA, in general, and omega-3 fatty acids, in particular. In Study I, the omega-3 fatty acid DPA and relative EPA increased (see figure 11.7a and b). DPA is a metabolite of EPA (see table 8.2) and is *not* present in fish-oil concentrate. In Study II, with the addition of fish-oil concentrate, DPA was unchanged, whereas relative EPA further increased (from about 15% before, to 25% [Study I] and 38% [Study II]) (see figure 11.7a).

The ratio of omega-6 over omega-3 is another way to describe metabolism of the two EFA series (see table 8.2). That ratio decreased in both studies as the plasma omega-3 EPA level increased (figure 11.7c). In Study I this was an effect of the antioxidant nutratherapy, and in Study II a combination of antioxidants and the fish-oil concentrate.

It is plausible that a more satisfactory antioxidant supplement in Study II would further enhance the omega-3 and depress the omega-6 fatty acids (see figure 11.7d).

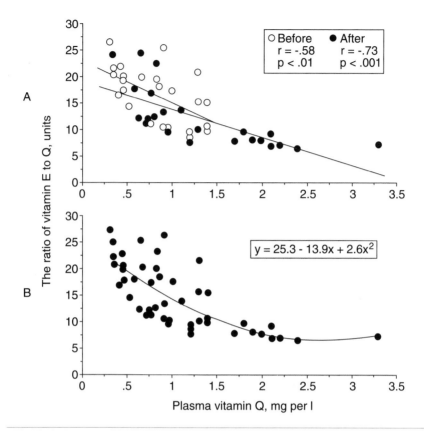

Figure 11.6 (a) The individual relationships before and after nutratherapy (Study I) for the ratio of vitamin E to Q versus plasma vitamin Q [Karlsson 1995]. (b) The same data as in (a), but pooled and assuming a polynomial relationship between the ratio of vitamin E to Q and plasma vitamin Q.

As discussed earlier (see figure 9.8), there is a leveling off in plasma EPA at approximately 60 mg per L for an EPA intake in excess of 1.0 g per day. This leveling off was indicative of saturation-like conditions that were present for plasma-borne EPA. An intake of 1.0 g per day of EPA is equivalent to a fish-oil concentrate supplement of 4 g per day, as in Study II. It is tempting to speculate that a higher EPA intake (i.e., a point at the steeper section of the relative EPA curve), could be more beneficial to suppress arachidonic acid.

To obtain a satisfactory suppressed omega-6 pattern, in general, and for arachidonic acid, in particular, it seems reasonable to conclude that antioxidant nutratherapy is not enough. The addition of omega-3 fatty acids seems to be nutraceutically mandatory provided that extra antioxidant support is administered.

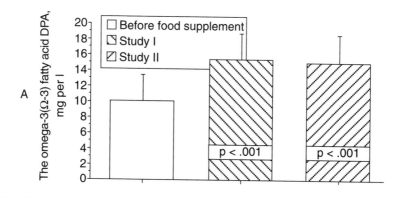

Figure 11.7a The omega-3 fatty acid DPA (see table 8.2) before and after the two separate food supplement regimens: Study I (only antioxidants) and Study II (antioxidants plus fish-oil concentrate) [Karlsson 1995]. Fish-oil concentrate contains only the omega-3 fatty acids EPA and DHA (see table 2.3). DPA is an endogenous metabolite of EPA. The addition of fish-oil concentrate had no further nutraceutical effect on DPA (compare plasma EPA regulation following EPA administration, chapter 8 and figure 9.8).

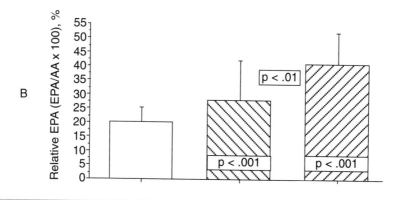

Figure 11.7b The relative EPA (EPA/AA x 100) before and after Study I and II [Karlsson 1995]. The addition of fish-oil concentrate further enhanced the ratio and subsequently nutraceutically suppressed the undesired omega-6 fatty acid arachidonic acid (AA), the precursor of the most aggressive prostanoids (prostaglandins).

Even among the XC skiers, the antioxidants had a nutraceutical significance and were found to improve physical health. For those XC skiers who had more than 1.1 mg vitamin Q per L plasma, the number of reported days missed from training due to disease was reduced by a

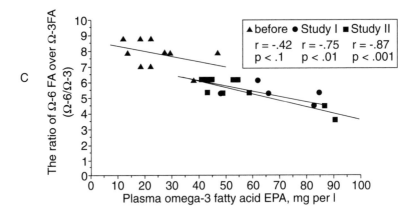

Figure 11.7c The individual relationships between the ratio of omega-6 to omega-3 fatty acids (the Ω-6/Ω-3 ratio) in plasma and plasma levels of the omega-3 fatty acid EPA (present in fish-oil concentrate) before the studies, in Study I, and in Study II, respectively [Karlsson 1995].

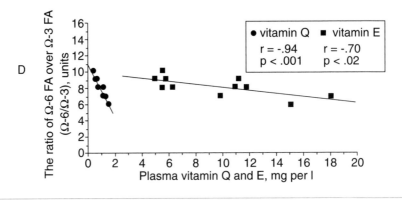

Figure 11.7d Both vitamins Q and E were at baseline (before nutratherapy), and are negatively related to the Ω-6/Ω-3 ratio [Karlsson 1995].

fourth or a fifth during the major competition months as compared to those who had less than 0.8 mg per L (see figure 11.8a). Mean values in healthy humans are 0.8 mg per L (i.e., values greater than or equal to 1.1 mg per L), indicating some form of nutratherapy. For those XC skiers who had plasma values equal to or more than 0.8 mg per L, there was a linear increase in the number of competition days versus their plasma vitamin Q levels (see figure 11.8b).

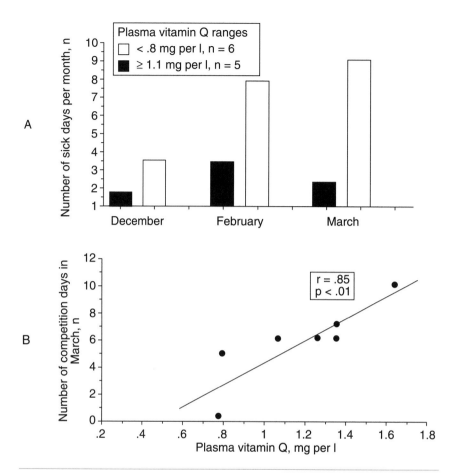

Figure 11.8   (a) Reported number of sick days during the competition season of 1992-93 (Study I) for those who undertook and were successful with their nutraceutical program (i.e., a plasma vitamin Q of ≥1.1 mg per L) as compared to those who either did not participate or failed [Karlsson 1995]. (b) Number of competition days in March 1993 for those with a nutraceutical effect on their plasma vitamin Q level (panel a).

Summary

Placebo-controlled studies prove the nutraceutical efficacy of both vitamin E and omega-3 fatty acids nutratherapy with respect to the immune system.

Nutratherapy with antioxidants, in general, elevates exercise performance capacity.

Data are lacking regarding the placebo-controlled efficacy of vitamin Q nutratherapy on the immune system. This is understandable, as vitamin Q's scientific history is much shorter than vitamin E's, and vitamin Q has been studied for only a short time.

In general, in open nutratherapy studies, with either matched controls or with the subjects as their own controls, lipophilic antioxidants have been found to reduce overuse injury and to improve exercise performance capacity.

In the controlled and open nutratherapy studies as well, it has been possible to relate performance changes to corresponding muscle and plasma antioxidant levels. This has been the basis on which to identify and establish clinical routines and set laboratory test standards for plasma vitamin contents, therapeutic levels, and an adequate nutratherapy according to the clinically applied substitution therapy concept.

Open nutratherapy studies with pharmaceutical-grade products (good clinical trial practice studies) have shown how food supplements with vitamins Q and E, with or without the addition of omega-3 fatty acids, improve the individual's health. As a result, fewer days are missed due to disease or injury, and the number of training or competition days is therefore increased.

12

Nutratherapy, Dose Response, and Side Effects

Prospective studies of vitamin E nutratherapy in connection with preventive cardiology and oncology have been reported in scientific journals [Rimm et al. 1992; Stampfer et al. 1992; BCCPSG 1994] and cited in the mass media. As a result of these studies, vitamin E nutratherapy has now received general acceptance as an adjuvant treatment for cardiovascular diseases. With regard to cancer prevention, however, a recently conducted study on Finnish smokers showed no reduction in cancer development with vitamin E and beta-(ß-)carotene nutratherapy [BCCPSG 1994].

Nutratherapies and Their Pharmacokinetics

Considering the recent interest shown in vitamin E nutratherapy by the medical community [Taylor, Dawsey, and Albanes 1990], it is surprising that data on the pharmacokinetics of vitamin E in health and disease are so scarce. An even greater lack of such data exists regarding vitamin Q nutratherapy, as from a scientific point of view vitamin Q is 50 years younger.

Nutritionists have paid little or no attention to the field of nutrient pharmacokinetics, which is astonishing because terms such as the Mediterranean Diet [Keys, Aravanis, and Blackburn 1967] and the french paradox [Frankel et al. 1993] have been well accepted and applied in preventive medicine, particularly for their prophylactic impact on cardiovascular health [Hertog et al. 1993]. Research into major nutrient absorption routes in the digestive system via different food sources and food preparations for vitamins Q and E is practically nonexistent. This has left the field open for speculation, extrapolation, or unproven hypothesis. As a result of this negligence among nutritionists, there are very few studies in functions such as the bioavailability of nutrients.

Recent epidemiological and cardiological studies of different nutratherapy programs with similar results, however, have caused scientists to question such nutraceutical concepts as the Recommended Dietary Allowance (RDA) and their implications [NRC, 1989a and b].

Evidently, the lipophilic nutrients—vitamins Q and E—cannot be directly absorbed through a well-defined receptor or channel in the digestive system. Nutrients such as vitamin F_1 are water soluble as fatty acids after lipolysis and can be absorbed through the mucosa of the small intestines. Fat can also be absorbed from the viscera as triglycerides, provided that they are in the stage of a microemulation (micelles). It has been suggested that a major absorption route for the lipophilic and lipoidic vitamins could be through these microscopic fat droplets as dissolved lipophilic compounds. These micelles appear later in blood and lymph as the chylomicrons [Traber, Cohn, and Muller 1993; Karlsson, Diamant, Theorell, Johansen, et al. 1993].

Comparative studies have also added indirect support to this notion because vitamin Q and E absorptions are improved if the vitamins are administered together with meals rather than between meals [Bogentoft et al. 1991]. The normal fat content in our diet is both a solvent and a transport vehicle of these normally membrane-bound lipophilic nutrients. Lipid malabsorption also reduces vitamin E uptake [Kowdley et al. 1992]. The digesting and mixing processes in the ventricle provide the opportunity for these nutrients to blend with and be dissolved in the normal dietary fat content. As discussed in chapter 2 (see figure 2.4), distribution, uptake, and plasma vitamin E values are related to the fat content in our diet. Nutrient deficiencies will develop if fat intake falls below 30% to 35% energy (E%F) or if other factors causing malabsorption are present [Muller, Harris, and Lloyd 1974]. In fact, some low-fat, long-term diets designed for endurance athletes have caused symptoms of malnourishment (see chapter 2).

Antioxidants and their protective faculties are partly directed toward the polyunsaturated fatty acids, in general, and the essential fatty acids, in particular. Of the different EFA series, the western-style diet favors the omega-6 fatty acids over the omega-3 fatty acids, which most investigators believe are insufficient in the western diet [Drevon 1992; Hay, Durber, and Saynor 1982; Ziboh et al. 1986]. Because of this insufficiency, many nutratherapy programs for elite athletes are based on a combination of antioxidants and fish-oil concentrate (see chapter 11 and figure 11.7).

Antioxidant Supplements and Tissue Changes

The individual effect of vitamin Q nutratherapy on muscle and plasma vitamin Q values varies considerably [Hay, Durber, and Saynor 1982; Ziboh et al. 1986]. One significant factor is the initial or baseline levels (i.e., the levels before start of a supplement program). It seems reasonable to assume that for a particular individual, these levels are determined by turnover factors, such as endogenous synthesis and breakdown. Some kind of equilibrium, or "flow," has been established in and between different compartments (see figure 12.1a-c).

A nutratherapy based on vitamin Q (100 mg per day for 6 weeks) caused muscle and plasma levels to increase from 0.08 to 0.23 g per kg and from 0.7 to 1.3 mg per L. The final plasma values, in contrast to the muscle values, were on an individual basis linearly related to baseline ("run-in") data (see figure 12.2a and b) [Karlsson, Diamant, Theorell, et al. 1991]. Those who had low plasma levels at the start (i.e., high muscle values; see figure 12.2a), had no or small plasma increases.

The same response pattern is also true for vitamin E nutratherapy [Karlsson, Diamant, Theorell, Johansen, et al. 1993]. As stated in chapter 11 (see figure 11.1a), before and after nutratherapy, plasma vitamin E increases linearly with vitamin Q. There are, however, some qualitative differences that should be further investigated. One consideration might be the fact that the ratio of vitamin E to Q decreases the higher the plasma vitamin Q level (see figure 11.2).

This ratio decrease has been found repeatedly in our studies of vitamin Q and E nutratherapies in healthy volunteers and in patient groups. Vitamin Q has also been demonstrated to protect vitamin E, and it has a prosthetic-like activity toward vitamin E (see figure 9.4). On the other hand, vitamin E has no protective effect on vitamin Q. In addition, vitamin Q is a product of endogenous synthesis, whereas vitamin E is an exclusively exogenous and essential nutrient.

The ratio of vitamin E to Q ranges 15 to 30:1 at a nonsupplemented stage as compared to 5 to 10:1 after the vitamin Q or Q+E nutratherapies (see figures 11.2 and 11.6) have been administered. These antioxidants

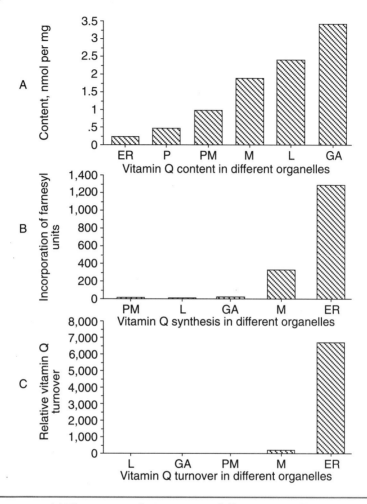

Figure 12.1   (a) The content of vitamin Q in different organelles of the cell (see figure 5.1) [Kalén et al. 1987]. (b) The relative incorporation rate of farnesyl (isoprene) units in the vitamin Q molecule [Kalén et al. 1987; Elmberger 1987]. (c) The estimated relative turnover of vitamin Q in different organelles [Kalén et al. 1987; Elmberger 1987].

are allocated to lipoidic structures. It has been demonstrated that the molar ratio for vitamin E to PUFA in membranes is 1:1,000 [Tappel 1980]. Moreover, vitamin E is associated with membrane zones characterized by high plasticity and fluidity as a result of their high contents of PUFA [Gomez-Fernandez, Villglain, and Aranda 1989]. PUFA need antioxidant protection. But even the vitamin Q and E molecules themselves enhance membrane fluidity [Yamamoto et al. 1986; Simon-Schnass and Korniszewski 1990].

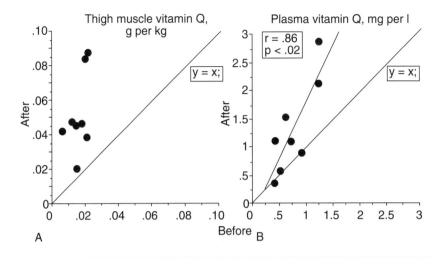

Figure 12.2 (a) Muscle vitamin Q content before and after 6 weeks of nutratherapy with 100 mg of vitamin Q per day. The specimens were obtained from the lateral portion of the thigh (*m vastus lateralis*) in physically active males [Karlsson, Diamant, Folkers, et al. 1991]. (b) Plasma vitamin Q level before and after 6 weeks of nutratherapy with 100 mg of vitamin Q per day for the same individuals as in panel (a) [Karlsson, Diamant, Folkers, et al. 1991].

Vitamin Q has the same nutraceutical property with respect to membrane antioxidant protection of PUFA and fluidity as vitamin E does [Crane 1990]. Hence, in a vitamin Q-depleted state, vitamin E may substitute for vitamin Q. The higher the ratio of vitamin E to Q, however, the less efficient vitamin E is toward Q.

Dose-Response Curves for Antioxidants

Therapeutic plasma vitamin Q and E levels have been set to 1.5 to 2.0 mg per L plasma (mean vitamin Q value for healthy controls is 0.8 mg per L) and 17 to 25 mg per L (mean vitamin E value is 12 mg per L) (see chapters 8 and 10) [Karlsson, Diamant, Theorell, Johansen, et al. 1993]. The basis has been plasma saturation-like conditions providing normal plasma lipids.

Dose-response studies of vitamin Q as a food supplement up to several hundred mg per day for a week have demonstrated that an intake of around 50 mg a day is needed to achieve a significant mean plasma increase over values corresponding to a well-balanced, mixed diet [Bogentoft et al. 1991; Karlsson, Diamant, Theorell, Johansen, et al. 1993]. Intake of more than 80 to 120 mg a day demonstrated leveling-off-like properties in average people, which could indicate plasma saturation-like conditions, as discussed in chapter 10.

It has already been stated that dietary intake of peroxidation-sensitive PUFA demands an increased antioxidant intake. This is most efficiently provided by vitamin E. From current studies, it could be estimated that 250 mg of vitamin E is needed per g of fish oil to avoid peroxidation and loss of vitamin E. This value is of the same order of magnitude as previous estimates by Dr. G. Block—400 mg per g PUFA [Murphy, Subar, and Block 1990]—and Dr. M. K. Horwitt—300 mg per g PUFA [Horwitt et al. 1988; Murphy, Subar, and Block 1990].

Vitamin C, unlike vitamins Q and E, is a water-soluble (hydrophilic) nutrient and represents a later stage in the antioxidant defense than the lipophilic antioxidants. By way of comparison, uptake mechanisms of this nutrient are discussed in this section. Plasma vitamin C level in healthy people, like all hydrophilic nutrients, is regulated by the uptake in the digestive system and urinary excretion.

Plasma vitamin C level varies according to the nutritional quality of the diet. A clinical average value in Scandinavia has arbitrarily been set to 6 mg per L (range, 1 to 7 mg per L). A daily vitamin C intake of 10 to 20 mg corresponds to a plasma level of 10 mg per L, 50 mg corresponds to 30 mg per L, and 600 mg corresponds to 100 mg per L [Geigy 1962]. This means that a leveling off occurs even for this hydrophilic antioxidant vitamin.

Vitamin C uptake is regulated by the absorption through the mucous membranes of the digestive system in the small intestines. It is estimated that almost all vitamin C is absorbed for daily intakes ranging 0 to 250 mg [Harris, Robinson, and Pauling 1973]. Administration of higher dosages will decrease the absorption rate and amount to only 50% or less at dosages of around 2 g per day and more.

Increased urinary excretion, together with reduced absorption, explain the plasma leveling-off-like conditions for this hydrophilic nutrient. In healthy people, urine excretion starts at a plasma level of about 15 mg per L [Geigy 1986].

A plasma vitamin C level corresponding to 15 mg per L or more means that vitamin C availability is not a limiting feature for pulmonary maximal oxygen uptake ($\dot{V}O_2$max) [Van der Beck 1985].

Most hydrophilic vitamins and nutrients are regulated in the same way as vitamin C. The rationale for such regulation mechanisms at both the absorption and excretion levels is most probably that antioxidants, at high tissue levels or when unbalanced, can become pro-oxidants [Wefers and Sies 1988]. The potential risk of the combination of vitamin C and the transit pool of free iron in the muscle has recently been stressed [Kanner, Harrel, and Hazan 1986; Jenkins and Goldfarb 1993; Jenkins and Halliwell 1994]. Theoretically, such a situation could also be valid for the lipophilic antioxidants such as vitamin E [Bowry and Stocker 1993]. The biological and clinical relevance of such a suspicion still needs to be proven.

In chapter 11, an antioxidant vitamin nutratherapy was shown to selectively enhance plasma levels of the omega-3 fatty acid series. It was also stated that a daily intake of 2 g of fish-oil concentrate further enhanced this condition and suppressed omega-6 fatty acids. This was particularly true of the omega-6 fatty acid arachidonic acid (AA), as described by the elevated relative EPA value (see table 8.2).

In chapter 8, figure 8.3, it was shown how a fish-oil concentrate intake of 2 g per day (equal to 1 g EPA per day) maximally elevated plasma EPA level (saturation-like conditions appeared). Additional fish-oil concentrate had no further enhancing effect on plasma EPA, whereas the relative EPA increased according to a positive curvilinear function, indicative of an accelerating AA suppression.

Many clinical studies on muscle and joint disorders (e.g., rheumatoid arthritis) and immune disorders (e.g., atopic eczema) have demonstrated the beneficial effects of AA suppression [Franson and Rosenthal 1985; Pruzanski et al. 1985; Rowley et al. 1984; Burton 1989; Morse et al. 1989; Oliwiecki et al. 1990; Bjørneboe et al. 1987; Søyland 1993; Søyland et al. 1993]. It seems reasonable to assume, then, that inflammatory processes in elite athletes (i.e., overuse injuries) can also be modulated as a result of an altered eicosanoid formation and synthesis promotion of less aggressive hormones by suppressing or avoiding the AA cascade (see chapter 9 and figure 9.6b).

Standardized or Individualized Nutratherapy?

The first and the prime objective, *before* a supplement program is begun, is to achieve satisfactory nourishment with a well-balanced diet. Such a diet will provide individuals with the highest possible amounts of the necessary macro- and micronutrients. A nutratherapy program can then be initiated on top of this.

In healthy people, therapeutic plasma vitamin Q and E levels (i.e., 1.5 to 2 mg and 17 to 25 mg per L; see table 12.1) , are obtained with a daily intake of

- 50 to 100 mg vitamin Q and
- 100 to 300 mg vitamin E.

It is recommended that this program be combined with

- 2 to 3 g fish-oil concentrate (vitamin F_1) and
- 1 to 2 g vitamin C.

This supplement program is designed for the average person with a normal physical conditioning level. A higher conditioning level demands extra needs. For example, for elite endurance athletes, the program can be doubled (200%). For elite soccer players and similar athletic groups, intakes of about 150% are recommended.

Micronutrients (e.g., selenium [Se]) will increase in plasma as a result of vitamin Q and E supplements alone. Increased absorption in the digestive system has been suggested as one possible explanation. (see chapter 9 and figure 9.9).

After a long-term supplement program (6 to 8 weeks), plasma levels of the most essential nutrients, such as vitamins Q, E, and F, and any measure of plasma lipids, should be checked.

For elite athletes and for those who either have repeated infectious diseases or overuse injuries, both initial ("run-in") data and data taken after 6 to 8 weeks should be obtained to individualize the therapy and to "titrate" an extra dosage, if necessary.

Side Effects of Vitamins Q, E, and F_1

In contrast to other lipophilic vitamins such as vitamins A and D, vitamins Q and E are free from severe side effects. Treatments of 300 to 400 mg vitamin Q [Greenberg and Frishman 1990] or 1 to 2 g vitamin E [Sies 1993; Meydani et al. 1990; Bendich and Machlin 1993] have been administered without any nutraceutical side effects.

In vitro studies have demonstrated that under certain conditions vitamin C with elevated tissue levels can act as a pro-oxidant [Ernster et al. 1992]. Recently, this has also been demonstrated with respect to vitamin E but not Q in laboratory experiments [Bowry and Stocker 1993]. The physiological and clinical significance of these experiments needs further investigation.

Nutratherapy based on vitamins Q and E is frequently combined with fish-oil concentrate containing several of the omega-3 fatty acid nutrients. These nutrients have a reducing or depressing effect on blood-clotting activity and platelet formation. Clotting mechanisms are essential to stop a wound's bleeding. It should therefore be noted that a prolonged bleeding time has been reported with respect to vitamin F_1 supplements [Leaf and Weber 1988]. Native Eskimos have an exclusive diet, very rich in fat fish or meat from animals living on these fish

Table 12.1 Normal Plasma Values in Healthy, Untreated Persons and the Corresponding Recommended Therapeutic Levels for Different Macro- and Micronutrients*

Nutrient	Generic name	Normal value	Therapeutic level
Vitamin Q	ubiquinone	≈ 0.8 mg per L	> 1.5 mg per L
Vitamin E	alpha-(α-)tocopherol	≈ 12	> 17
Vitamin F_1	omega-3 (Ω-3) fatty acids		
ALA		≈ 60	> 80
EPA		≈ 20	> 40
DPA		≈ 10	> 15
DHA		≈ 60	> 100
Vitamin F_2	omega-6 (Ω-6) fatty acids		
LA		≈ 500	< 400
GLA		≈ 10	< 5
DGLA		≈ 60	< 40
AA		≈ 140	< 100
Vitamin F ratios			
total Ω-6/Ω-3		≈ 10	< 5
EPA/AA × 100 (relative EPA)		≈ 15	> 25

(continued)

Table 12.1 *(continued)*

Nutrient	Generic name	Normal value	Therapeutic level
Nutrients as options			
Vitamin C	ascorbate	6	> 15
Beta-(B-)carotene	"pro-vitamin" A	2-3	**
Calcium (Ca)		100	**
Chromium (Cr)		0.03	**
Magnesium (Mg)		40	**
Manganese		0.04	**
Selenium (Se)		0.2	0.3
Zinc (Zn)		3	**

* These data are based on the assumption that the control subjects have normal plasma lipids and an average physical condition-ing level. *Elevated lipids* means that the listed normal and therapeutic figures are correspondingly elevated. *Lowered plasma lipids*, which endurance-trained subjects and patients treated with plasma-lipid-lowering drugs have, means that the listed values are decreased to make comparisons valid.

** Not defined

[Dyerberg 1986]. A longer bleeding time has been reported for native Eskimos than for urban Danes [Hay, Durber, and Saynor 1982]. Treatment with fish-oil concentrate has been applied extensively as an adjuvant therapy in clinical care [Draper 1993]. This treatment has frequently been combined with other drugs affecting circulation. Despite these treatment combinations, no bleeding side effects have been reported related to the pharmacologically active omega-3 fatty acids.

Whether the possible side effect of bleeding has any bearing on muscle ruptures, for example, has not been investigated. Therefore, a warning should be issued about combining fish-oil concentrate with salicylate or other nonsteroidal anti-inflammatory drugs (NSAID) known to affect blood clotting mechanisms, because of the possibility of additional bleeding.

Summary

Nutratherapy with vitamins Q and E enhances deposition of these antioxidants in both muscle and plasma. Individuals respond very differently, which is partly due to individual plasma baseline values (i.e., values after a mixed diet). Qualitative differences also exist before and after nutratherapy, which might be related to such intrinsic features as

- vitamin Q but not E as an endogenous product,
- the way in which vitamin Q acts (i.e., in a prosthetic-group-like fashion) toward vitamin E, and
- the ratio of vitamin E to Q being 15 to 30:1 in the nonsupplemented, as compared to 5 to 10:1 in the supplemented stage.

Bioavailability, pharmacokinetic, and dose-response studies are extremely rare for vitamin Q and E supplement or therapy programs. They are more abundant for nutrients such as omega-3 fatty acids (fish-oil concentrate).

Lipophilic compounds such as vitamins Q and E can only be absorbed by the mucosa of the intestines when they are dissolved in the microscopic, triglyceride-based, fat droplets—the micelles. They are absorbed as entities and further digested in the liver and reallocated to lipoprotein particles, including chylomicrons, and then released to the bloodstream. Other absorption receptors or mechanisms with respect to these nutrients are unknown.

Comparisons of vitamin Q and E nutratherapies regarding their administration at mealtime or between meals have shown that intake with meals is preferable because it enhances bioavailability.

Although individuals respond very differently to standardized nutratherapy programs, a standardized program has been suggested. This program is designed for the average person with normal physical conditioning. For elite endurance athletes, it can be doubled.

It is recommended that elite athletes combine their food supplement program with plasma determinations of the critical nutrients such as vitamins Q, E, and F. This recommendation is also valid for those who have frequent infectious diseases or overuse injuries and would like to improve their immune response.

13

Nutrients, Pharmaceutical Grade, and the Doping Issue

In preventive medicine, the beneficial effects of nutrients such as vitamins Q, E, and F_1 have been accepted as clinically proven [Packer and Fuchs 1993; Imagawa 1990; Mortensen 1993; Packer and Viguie 1989; Drevon 1992; Hofman-Bang, Rehnqvist, and Swedberg 1992]. Nutratherapy has, therefore, been applied as an add-on (adjuvant) therapy in clinical practice. This has moved the field of nutrition from the health food and grocery stores and the corresponding industries to the pharmacy and the pharmaceutical industry.

Supplement Products and Good Manufacturing Practice (GMP)

Nutratherapies are based on nutrients in concentrated forms and sometimes as artificial products; hence, they are comparable to an ethical drug therapy. For the safety of the consumer, as with ethical drugs, it is essential that the manufacture of nutraceuticals be carried out according to the regulations of Pharmaceutical Investigating Convention-Good Manufacturing Practice (PIC-GMP), as defined by the World Health Organization (WHO) [Lisook 1990]. This international code for products of pharmaceutical grade requires that they be manufactured with sources and procedures that guarantee the consumer safe and effective products.

The United States federal government has shown great insight and has taken major steps to protect the consumer. Federal authorities have also put forth a number of nutritional policy recommendations in recent years. Some of the major acts are as follows:

- 1991 National Nutritional Monitoring Act (NNMA)
- 1993 Nutritional Labeling and Education Act (NLEA)
- 1994 Dietary Supplement Health and Education Act (DSHEA)

Changes are being considered in the NNMA and DSHEA that would require the same product control protocols with regard to pharmaceutical, food, and nutrient products as holds for ethical drugs. The NLEA mandates the Food and Drug Administration (FDA) to approve health claims for food products and to re-examine the Recommended Daily Allowance (RDA) of critical nutrients.

The European Union (EU) is considering the same steps, but is presently hampered by internal debates on nutrient definition, historical backgrounds and perspectives among the EU members, discussions of RDA levels, and other issues.

The clinical use of nutrients as a treatment has motivated the pharmaceutical industry to develop nutraceutical products that meet the international PIC-GMP standards. These nutraceutical products can be prescribed and purchased as pharmaceuticals or as over-the-counter (OTC), nonprescription drugs. They are also sold as "food supplements" or "vitamins" in the health food store market, in accordance with the laws of most of the countries in Europe and North America. The laws for these food supplements are far less restrictive and demanding than the laws for pharmaceutical products. It is not surprising, then, that contamination, deterioration due to breakdown, and preparation with caffeine or ephedrine have frequently been reported and documented.

The regulation of nutraceutical products varies from country to country. In the United States, for example, vitamin F_2 (the omega-6 fatty acid series)-based nutraceutical supplements are forbidden as health food products, whereas vitamin Q is available in most states. In France, vitamin Q is not available as a health food product, but omega-6 fatty acid products are. In Japan and Italy, vitamin Q is available exclusively as a prescription drug.

Nutraceutical Therapy, Ergogenic Aids, and Sport Ethics

According to the International Olympic Committee (IOC) and its Medical Commission, "physiological substances in abnormal amounts . . . attaining an artificial and unfair increase of performance in competition" is synonymous with violating the doping rules [International Olympic Committee, and Medical Commission 1976, p. 1]. The potential effect of this statement on most nutraceuticals or other food supplement programs is obvious.

Over the years, different aspects of the wording of the IOC statement and its interpretation have sparked debate. In the spring of 1975, I participated in a meeting organized by the IOC Medical Commission in Porto Allegre, Brazil, where sport nutrition was on the agenda. The Italian representative had motioned that carbohydrate nutraceutical supplements and muscle glycogen loading was a doping offense and should, accordingly, be forbidden and punishable. The diet regimen to boost muscle glycogen was introduced in the late 1960s and rapidly won worldwide appreciation among such endurance athletes as marathoners, racing cyclists, cross-country skiers, and triathletes. No measures were taken by the Medical Commission regarding the Italian motion. This lack of action was interpreted by the international sports community as prejudicial by the IOC Medical Commission.

In 1991, in a consensus decision in Lausanne (International Scientific Consensus), the IOC stated that nutrient intake was sufficiently supplied by a well-balanced diet: "Supplements are not necessary . . . when eating a diet adequate in respect of quality and quantity . . . ." [Devlin and Williams 1991, p. 1].

As a representative of a Swedish pharmaceutical company, I had earlier (1983) requested a hearing on this matter with the IOC. The hearing was organized by a consultant to the IOC Medical Commission, the late Dr. Manfred Doniken. Another participant was the present chairman of the international society for sport physicians (Federation Internationale de Medicine Sportive, FIMS), Dr. Wilfred Hollman. It was concluded that nutrients consumed in concentrated and artificial forms could not be regarded as doping agents.

In the Summer Olympic Games of 1992, in Barcelona, Spain, the successful British sprinter Linford Christie announced at a press conference that he had applied nutratherapy and used the ergogenic aid creatine. He ascribed his gold medal in the 100-meter dash to this particular nutraceutical. This statement sparked a new debate about nutratherapy and the ethical aspects of its use in elite sports.

It has been claimed that creatine nutraceutical supplement belongs to a "grey zone" with respect to the doping regulation as instituted by the IOC. Expressions of that kind, however, are not scientifically based, and show no knowledge of current documentation concerning the biology of creatine and creatine phosphate synthesis in muscle exercise and diet [Harris, Söderlund, and Hultman 1992; Grennhaff et al. 1993].

Nutratherapy with antioxidant vitamins and other nutrients has been the object of the same kind of poor judgment and misinformed statements as those regarding creatine. Hence, when advising athletes about nutratherapies, individual physicians, trainers, and physical therapists sometimes act in a clandestine manner, as though dealing with illegal measures or hocus-pocus. This manner of conveying the information has given athletes the impression that nutraceuticals and nutratherapy are not amenable to open scientific or medical scrutiny.

Scientific and other organizations have recently been forced to act and present statements in the name of medical ethics, which theoretically violates the international regulations of the IOC. The most aggressive statement was made by the United States Olympic Committee and its Medicine Committee in the spring of 1994, which not only granted but also justified nutratherapy programs to prevent injury [Grandjean 1994]. Recently, Swedish sport authorities have adapted an approach to nutratherapy similar to that of the USOC [CPU 1995].

Elite Sport, "Drugs," and the Doping Rules

A nutratherapy program is a true pharmaceutical intervention. In accordance with this, legislation in most European countries defines a supplement of concentrated nutrients in excess of RDA as a drug treatment. In Denmark, for example, laws for "strong vitamins" are as restrictive as laws for ethical drugs. Many countries, though, including Sweden, do not yet enforce such tight drug legislation on nutraceuticals.

Theoretically, in accordance with the present rules, a nutraceutical program could be in violation of the doping regulation. The biological and medical rationale for nutratherapy is to enhance protection against injury from physical activities and to maintain health. Such a program can, of course, indirectly affect and increase physical performance. But such an improvement cannot be evaluated differently from that seen with adaptation to increased muscle activity and training.

The doping regulation must be written in such a way that nutratherapy in fitness and elite sports cannot be questioned as an unethical measure. Sport ethics and moral debate must be in line with our present biological and medical knowledge and their progress. Sports, correctly executed, can be a health-promoting activity, for both the fitness-motivated person and the elite athlete.

It must be said again, as a final statement, that a nutratherapy program cannot substitute for a nutritious, well-balanced diet. However, based on such a diet, it can improve the well-being of athletes and can prevent diseases and overuse injury.

SECTION

IV

Conclusions

1. Cell metabolism is not only a matter of fermentation and respiration but also is the basis of free radical formation.
2. Free radical formation occurs in all living cells and, in most cases, is synonymous with extremely reactive metabolites whose formation rate of oxygen utilization is on the order of 3% to 15%.
3. If not scavenged, these reactive species can react in an uncontrolled manner, resulting in derangement and injury to the cell, thus threatening the cell. But radicals are also necessary for forming half-manufactures and by-products of the cell and its regeneration.
4. Radical scavenging is a biochemical reaction where the pro-oxidant characteristics of the radical are reduced by means of reduced compounds either with or without enzymes; hence, the terms *antioxidants* and *antioxidant enzymes*.
5. As most metabolic processes occur in lipid layers and membranes of the cell, the primary antioxidant activity is carried out by the lipid-soluble (lipophilic) antioxidant defense.
6. The second antioxidant defense line is in the water phase of the cytosol and is based on water-soluble (hydrophilic) antioxidants.
7. The sole duty of many vitamin nutrients is to act as antioxidants, whereas ubiquinone (vitamin Q), the most central and crucial antioxidant, also acts as a coenzyme (coenzyme Q, or CoQ_{10}) in the mitochondria between the citric acid cycle and the respiratory chain.
8. As a coenzyme, ubiquinone feeds electrons from mitochondria into the antioxidant systems and keeps the rest of the antioxidant mechanism reduced and on the alert.

9. Vitamin Q is also required as a catalyst for the antioxidant role of vitamin E, which is quantitatively the most significant antioxidant in lipids.

10. To unload vitamin E and expand the antioxidant activity into the water phase, vitamin C has a special function with regard to vitamin E in the interphase between lipids and water. This can also be seen as a catalyzing system to exploit the rest of the vitamin C pool, vitamin P (bioflavonoids), antioxidant metabolites, glutathione, and the different antioxidant enzyme systems. The antioxidants involved in this scheme are referred to as the vitamin Q-E-C cycle.

11. Vitamins E and C are essential nutrients, which means that they must be present in sufficient amounts in the diet. Ubiquinone, however, can be synthesized in all organs and cell systems investigated. The formation rate evidently does not match the turnover of ubiquinone in some situations and the diet content of vitamin Q will be significant. That is the rationale for calling ubiquinone vitamin Q.

12. An uncontrolled radical formation will threaten cell constituents. They constitute cell compounds such as lipids and, predominantly, the fraction of polyunsaturated fatty acids (PUFA), chondroitin structures including hyaluronic acids, genome structures as represented by DNA and RNA, and so on. Peroxidation of lipids can become a self-generating phenomenon known as the *cascade reaction*. The cascade reaction will severely damage the lipid layers, including cell membranes, and cause cell decomposition and death, invasion of leukocytes, and so on. An inflammatory process—overuse injury—then develops.

13. PUFA exist in different groups or series, of which two are essential nutrients: omega-3 (Ω-3) and omega-6 fatty acids (Ω-6). In the 1920s they were given the name vitamin F (vitamins A through E had already been assigned). These PUFA series are also referred to as essential fatty acids (EFA).

14. Our present urban western diet favors more consumption of omega-6 fatty acids (vitamin F_2) than omega-3 fatty acids (vitamin F_1). The ratio is approximately 30:1 as compared to 5:1 in coastal populations with a mixed diet, which includes fish and predators of fish. It is estimated that the ratio 10,000 to 50,000 years ago was 1:1. The present diet favors the enhanced intake of one of the omega-6 fatty acids: arachidonic acid (AA). AA is the precursor of the most aggressive and painful prostanoid (prostaglandin) series.The corresponding prostanoids appear in severe inflammatory processes. This is referred to as the arachidonic acid cascade.

15. Not only is our vitamin F_1 intake lower, but also on a molar basis it is more susceptible to peroxidation than is vitamin F_2. There are ample epidemiological data to support the notion that either an enhanced fish intake or a food supplement (nutratherapy) with either fish liver oil (from the liver) or fish oil (from muscle tissue) can change this pattern and improve cardiovascular diseases, arthritis, skin disorders, and other signs of a deficient immune response and inflammatory processes.

16. Antioxidant vitamins have a protective role with respect to peroxidation of the EFA, particularly vitamin F_1.

17. Controlled nutratherapy studies in healthy people and in different patient groups have shown that vitamins Q and E have clinical effects in accordance with their antioxidant functions. Nutratherapy with vitamin F_1 and its membrane-stabilizing effect, which, among other things, improves the immune response, is also well demonstrated.

18. Cross-sectional studies in fitness and elite athletes in different sports have demonstrated depleted or exhausted antioxidant values in muscle and plasma levels as well as distorted vitamin F values: lower vitamin F_1, and normal to above-normal F_2 levels.

19. Substitution studies have elevated muscle and plasma antioxidant vitamin levels. Simultaneously, vitamin F levels have improved—vitamin F_1 levels have increased, and, in relative terms, F_2 levels have gotten lower. The addition of vitamin F_1 in the substitution nutratherapy further improved F_1 levels both in absolute and relative terms.

20. Controlled studies in fitness and elite athletes have repeatedly shown that vitamin Q and E nutratherapy improves physical performance and reduces signs of radical injury and fatty acid peroxidation.

21. No serious side effects have been observed in either healthy volunteers or patients after prolonged nutratherapy with the antioxidant vitamins Q and E or vitamin F_1.

References

Adelman, R., R.L. Saul, and B.N. Ames. 1988. Oxidative damage and DNA: Relation to species metabolic rate and life span. *Proc. Nat. Acad. Sci. U.S.A.* 85: 2706-2708.

Agnevik, G., J. Karlsson, L. Hermansen, and B. Saltin. 1967. Energy demands during running. In *II. Internationales Seminar fur Ergometrie*. Berlin: Ergon Verlag.

Ahlborg, G., and M. Jensen-Urstad. 1991. Metabolism in exercising arm vs. leg muscle. *Clin. Physiol.* 11: 459-468.

Åkermark, C., I. Jacobs, M. Rasmusson, and J. Karlsson. 1996. Performance in ice hockey and relation to muscle morphology and muscle glycogen content. *Int. J. Sports Nutr.* 6: 272-284.

Alessio, H.M. 1993. Exercise-induced oxidative stress. *Med. Sci. Sports Exerc.* 25: 218-224.

Alessio, H., M. Cutler, and R.G. Cutler. 1990. Evidence that DNA damage and repair cycle activity increases following a marathon race [Abstract]. *Med. Sci. Sports Exerc.* 22: 751.

Ames, B.N. 1989. Endogenous oxidative DNA damage, aging and cancer. *Free Radic. Res. Commun.* 7: 121-128.

Andersen, P. 1975. Capillary density in skeletal muscle of man. *Acta Physiol. Scand.* 95: 203-205.

Appelkvist, E.-L., A. Kalén, and G. Dallner. 1991. Biosynthesis and regulation of coenzyme Q. In *Biomedical and clinical aspects of coenzyme Q*, ed. K. Folkers, G.P. Littarru, and T. Yamagami. Amsterdam: Elsevier.

Armstrong, R.B., P.D. Gollnick, K. Piehl, B. Saltin, and C.W. Saubert, IV. 1972. Enzyme activity and fiber composition in human skeletal muscle. *Acta Physiol. Scand.* 84 (Suppl. A): 34-36.

Artola, A., J.L. Alio, J.L. Bellot, and J.M. Ruiz. 1993. Lipid peroxidation in the iris and its protection by means of viscoelastic substances (sodium hyaloronate and hydroxypropylmethylcellulose). *Ophthalmic Res.* 25: 172-176.

Asp, K.-Å., Å. Bruce, and L. Hambreus. 1994. Vitaminer och antoixidantia—behövs kosttillskott? *Scand. J. Nutr.* 38: 1-124.

Balsom, P.D., J.Y. Seger, B. Sjödin, and B. Ekblom. 1992. Maximal-intensity intermittent exercise: Effect of recovery duration. *Int. J. Sports Med.* 13: 528-533.

Bassenge, E. 1992. Clinical relevance of endothelium-derived relaxing factor (EDRF). *Br. J. Clin. Pharmacol.* 34 (Suppl. 1): 375-425.

BCCPSG (Beta Carotene Cancer Prevention Study Group). 1994. The effect of vitamin E and beta carotene on the incidence of lung cancer and other cancers in male smokers. *N. Engl. J. Med.* 14: 1029-1035.

Bélichard, P., D. Pruneau, and A. Zhiri. 1993. Effect of a long-term treatment with lovastatin or fenofibrate on hepatic and cardiac ubiquinone levels in cardiomyopathic hamster. *Biochim. Biophys. Acta* 1169: 98-102.

Bendich, A. 1993. Vitamin E and human immune functions. *Nutrition and immunology.* New York: Plenum Press.

Bendich, A., and L.J. Machlin. 1993. The safety of oral intake of vitamin E: Data from clinical studies from 1986 to 1991. In *Vitamin E in health and disease,* ed. L. Packer and J. Fuchs. New York: Marcel Dekker.

Berg, U., A. Thorstensson, B. Sjödin, B. Hultén, K. Piehl, and J. Karlsson. 1978. Maximal oxygen uptake and muscle fiber types in trained and untrained humans. *Med. Sci. Sports* 10: 151-154.

Berg Schmidt, E., J.O. Pedersen, K. Varming, E. Ernst, C. Jersild, N. Grunnet, and J. Dyerberg. 1991. n-3 fatty acids and leukocyte chemotaxis. *Arterioscler. Thromb.* 11: 429-435.

Berg Schmidt, E., K. Varming, E. Ernst, P. Madsen, and J. Dyerberg. 1990. Dose-response studies on the effects of n-3 polyunsaturated fatty acids on lipids and haemostasis. *Thromb. Haemost.* 63: 1-5.

Bergström, J., and E. Hultman. 1966. Muscle glycogen synthesis after exercise: An enhancing factor localized to the muscle cells in man. *Nature* 210: 309-310.

Berne, R.M. 1980. The role of adenosine in the regulation of coronary blood flow. *Circ. Res.* 47: 807-813.

Beving, H.F., B. Tedner, and G. Eriksson. 1991. Study of the electrical impedance of blood from house painters. Submitted for publication.

Beving, H.F., G. Petrén, and O. Vesterberg. 1990. Increased isotransferrin ratio and reduced erythrocyte and platelet volumes in blood from thermoplastic industry workers. *Ann. Occup. Hyg.* 34: 391-397.

Beyer, R.E., and L. Ernster. 1990. The antioxidant role of coenzyme Q. In *Highlights in ubiquinone research,* ed. G. Lenaz, O. Barnabei, A. Rabbi, and M. Battino. London: Taylor and Francis.

Beyer, R.E., K. Nordenbrand, and L. Ernster. 1986. The role of coenzyme Q as a mitochondrial antioxidant: A short review. In *Biomedical and clinical aspects of coenzyme Q,* ed. K. Folkers and Y. Yamamyra. Amsterdam: Elsevier.

Biörck, G. 1949. On myoglobin and its occurence in man. *Acta Med. Scand.* 226 (Suppl.):1-145.

Bjørneboe, A., E. Søyland, G.-E. Bjøneboe, G. Rajka, and C.A. Drevon. 1987. Effect of dietary supplementation with eicosapentaenoic acid in the treatment of atopic dermatitis. *Br. J. Dermatol.* 117: 463-469.

Blake, D.R., J. Unsworth, J.M. Outhwaite, C.J. Morris, P. Merry, B.L. Kidd, R. Ballard, and L. Gray. 1989. Hypoxic-reperfusion injury in the inflamed human joint. *Lancet* 8633: 289-293.

Blix, G. 1965. A study on the relation between total calories and single nutrients in food. *Acta Soc. Med. Upsalien.* 70: 17-129.

Bogentoft, C., P.-O. Edlund, B. Olsson, L. Widlund, and K. Westensen. 1991. Biopharmaceutical aspects of intravenous and oral administration of coenzyme Q. In *Biomedical and clinical aspects of coenzyme Q*, ed. K. Folkers and Y. Yamamyra. Amsterdam: Elsevier.

Boveris, A. 1977. Mitochondrial production of superoxide radical and hydrogen peroxide. *Adv. Exp. Med. Biol.* 78: 67-82.

Boveris, A., E. Cadenas, R. Reiter, M. Filipkowski, Y. Nakase, and B. Chance. 1980. Organ chemiluminescence: Noninvasive assay for oxidative radical reactions. *Proc. Natl. Acad. Sci. U.S.A.* 77: 347-351.

Bowry, V.W., and R. Stocker. 1993. Tocopherol-mediated peroxidation: The prooxidant effect of vitamin E on the radical-initiated oxidation of human low-density lipoprotein. *J. Am. Chem. Soc.* 115: 6029-6044.

Braunwald, E., and R. Kloner. 1985. Myocardial reperfusion: A double-edged sword? *J. Clin. Invest.* 76: 1713-1719.

Brown, M.S., and J.L. Goldstein. 1983. Lipoprotein metabolism in the macrophage: Implications for cholesterol deposition in atherosclerosis. *Annu. Rev. Biochem.* 52: 223-261.

Burton, G.W., and K.U. Ingold. 1981. Autooxidation of biological molecules. I. Antioxidant activity of vitamin E and related chain-breaking phenolic antioxidants in vitro. *J. Am. Chem. Soc.* 103: 6472-6477.

———. 1989. Vitamin E as an *in vitro* and *in vivo* antioxidant. *Ann. N.Y. Acad. Sci.* 570: 7-22.

———. 1993. Biokinetics of vitamin E using deuterated tocopherols. In *Vitamin E in health and disease*, ed. L. Packer and J. Fuchs. New York: Marcel Dekker.

Burton, G.W., A. Joyce, and K.U. Ingold. 1983. Is vitamin E the only lipid-soluble, chain-breaking antioxidant in human blood plasma and erythrocyte membranes? *Arch. Biochem. Biophys.* 221: 281-290.

Burton, J.L. 1989. Dietary fatty acids and inflammatory skin disease. *Lancet.* 8628: 27-30.

Buzina, R., and K. Suboticanec. 1985. Vitamin C and physical working capacity. *Int. J. Vitam. Nutr. Res.* 27 (Suppl.): 157-164.

Cabrini, L., B. Paasquali, B. Tadolini, A.M. Sechi, and L. Landi. 1986. Antioxidant behaviour of ubiquinone and ß-carotene incorporated in model membranes. *Free Rad. Res. Commun.* 2: 85-92.

Chance, B. 1947. An intermediate compound in the catalytic-hydrogen peroxide reaction. *Acta Chem. Scand.* 1: 236-267.

Chance, B., and E. Pring. 1968. Logic in the design of the respiratory chain. In *Biochemie des Sauerstoffs*, ed. B. Hess and H. Staudinger, 102-126. Berlin: Springer-Verlag.

Chance, B., B. Schoener, and F. Chindler. 1964. The intracellular oxidation-reduction state. In *Oxygen in the animal organism*, ed. F. Dickens and E. Neil, 367-388. London: Pergamon Press.

Choudhury, N.A., S. Sakaguchi, K. Koyano, A.F. Matin, and H. Muro. 1991. Free radical injury in skeletal muscle ischemia and reperfusion. *J. Surg. Res.* 51: 392-398.

Christensen, E.H., and O. Hansen. 1939a. Arbeitsfähigkeit und Ernährung. *Skand. Arch. Physiol.* 81: 160-171.

———. 1939b. Hypoglykämi, Arbeitsfähigkeit und Ernährung. *Skand. Arch. Physiol.* 81: 172-179.

————. 1939c. Untersuchungen über die Verbrennungsvorgänge bei langdauernder, schwerer Muskelkarbeit. *Skand. Arch. Physiol.* 81: 152-159.

Clausen, J. 1991. The influence of selenium and vitamin E on the enhanced respiratory reaction in smokers. *Biol. Trace Elem. Res.* 31: 281-291.

CPU (Centrum för Prestations Utveckling). 1995. Kost & kosttillskott. *Guldkorn.* 10: 1-5.

Crane, F., Y. Hatefi, R.L. Lester, and C. Widmer. 1957. Isolation of quinone from beef heart mitochondria. *Biochim. Biophys. Acta* 25: 220-221.

Crane, F.L. 1990. Development of concepts for the role of ubiquinones in biological membranes. In *Highlights in ubiquinone research*, ed. G. Lenaz, O. Barnabei, A. Rabbi, and M. Battino. London: Taylor and Francis.

Crane, F.L., I.L. Sun, E. Sun, and D.J. Morré. 1991a. Antioxidant functions of coenzyme Q: Some biochemical and pathophysiological implications. In *Biomedical and clinical aspects of coenzyme Q*, ed. K. Folkers, G.P. Littarru, and T. Yamagami. Amsterdam: Elsevier.

Crane, F., I.L. Sun, E.E. Sun, and D.J. Morré. 1991b. Cell growth stimulation by plasma membrane electron transport. *FASEB J.* 5: A1624.

Crane, F.L., I.L. Sun, and E.E. Sun. 1993. The essential functions of coenzyme Q. *Clin. Invest.* 71(supp.): 55-59.

Cui, J., M. Garle, P. Eneroth, and I. Bjorkhem. 1994. What do commercial ginseng preparations contain? *Lancet* 344: 134.

Cunnane, S.C., S.-Y. Ho, P. Dore-Duffy, K.R. Ells, and D.F. Horrobin. 1989. Essential fatty acid and lipid profiles in plasma and erythrocytes in patients with multiple sclerosis[1-4]. *Am. J. Clin. Nutr.* 50: 801-806.

Dallner, G. 1994. Coenzyme Q. *Scand. J. Nutr.* 38: 84-86.

Davies, K.J.A., A.T. Quintanidha, G.A. Brooks, and L. Packer. 1982. Free radicals and tissue damage produced by exercise. *Biochem. Biophys. Res. Commun.* 107: 1198-1205.

Deby, C., and R. Goutier. 1990. New perspectives on the biochemistry of superoxide anion and efficiency of superoxide dismutase. *Biochem. Pharmacol.* 39: 399-405.

Del Maestro, R.F., H.H. Thaw, J. Björk, M. Planker, and K.-E. Arfors. 1980. Free radicals as mediators of tissue injury. *Acta Physiol. Scand.* 492 (Suppl.): 43-57.

Demopoulos, H.B., J.P. Santomier, M.L. Seligman, and L. Pietronigro. 1984. Free radical pathology: Rationale and toxicology of antioxidants and other supplements in sports medicine and exercise science. In *Sport health and nutrition*, ed. F.I. Katch. Champaign, IL: Human Kinetics.

Devlin, J.T., and C. Williams. 1991. Foods, nutrition and sports performance: Proceedings of an International Scientific Consensus held 4-6 February 1991, Lausanne. *J. Sports Sci.* 9: 1-152.

Dillard, C.J., R.E. Litov, W.M. Savin, E.E. Dumelin, and A.L. Tappel. 1978. Effects of exercise, vitamin E, and ozone on pulmonary function and lipid peroxidation. *J. Appl. Physiol.* 45: 927-932.

di Prampero, P.E. 1981. Energetics of muscular exercise. *Rev. Physiol. Biochem. Pharmacol.* 89: 143-222.

Draper, H.H. 1993. Interrelationships of vitamin E with other nutrients. In *Vitamin E in health and disease*, ed. L. Packer and J. Fuchs. New York: Marcel Dekker.

Drevon, C.A. 1992. Marine oils and their effects. *Nutr. Rev.* 50: 29-36.

Duthie, G.G. 1993. Antioxidant status in smokers. In *Vitamin E in health and disease*, ed. L. Packer and J. Fuchs. New York: Marcel Dekker.

Dyerberg, J. 1986. Linoleate-derived polyunsaturated fatty acids and prevention of atherosclerosis. *Nutr. Rev.* 44: 125-133.

Dyerberg, J., H.O. Bank, E. Stoffersen, E.S. Moncada, and J.R. Vane. 1978. Eicosapentaenoic acid and prevention of thrombosis and atherosclerosis. *Lancet* 2: 117-119.

Eaton, S.B., and M. Konner. 1985. Paleolithic nutrition: A consideration of its nature and current implications. *N. Engl. J. Med.* 312: 383-389.

Edlund, P.-O. 1988. Determination of coenzyme Q_{10}, α-tocopherol and cholesterol in biological samples by coupled-column liquid chromatography with coulometric and ultraviolet detection. *J. Chromatogr. A* 425: 87-97.

Ekblom, B. 1994. *Football (Soccer)*. London: Blackwell Scientific.

Ekholm, R., J. Kerstell, R. Olsson, C.-M. Rudenstam, and A. Svanborg. 1968. Morphological and biochemical studies of dog heart mitochondria after short periods of ischemia. *Am. J. Cardiol.* 22: 312-318.

el-Hage, S., and S.M. Singh. 1990. Temporal expression of genes encoding free radical-metabolizing enzymes is associated with higher mRNA levels during in utero development in mice. *Dev. Genet.* 11: 149-159.

Engström-Laurent, A., and R. Häägren. 1987. Circulating hyaluronic acid varies with physical activity in health and rheumatoid arthritis. *Arthritis Rheum.* 30: 1333-1338.

Ernster, L. 1986. Oxygen as an environmental poison. *Chemica Scripta.* 26: 525-534.

Ernster, L., and R.E. Beyer. 1991. Antioxidant functions of coenzyme Q: Some biochemical and pathophysiological implications. In *Biomedical and clinical aspects of coenzyme Q*, ed. K. Folkers, G.P. Littarru, and T. Yamagami. Amsterdam: Elsevier.

Ernster, L., P. Forsmark, and K. Nordenbrand. 1992. The mode of action of lipid-soluble antioxidants in biological membranes: Relationship between the effects of ubiquinol and vitamin E as inhibitors of lipid peroxidation in submitochondrial particles. *BioFactors.* 3: 241-248.

Ernster, L., and P. Forsmark-Andrée. 1993. Ubiquinol - an endogenous antioxidant in aerobic organisms. *Clin. Invest. Med.* 71(supp.): 60-65.

Ernster, L., and C.P. Lee. 1990. Thirty years of coenzyme Q: Some biochemical and pathophysiological implications. In *Bioenergetics: Molecular biology, biochemistry and pathology*, ed. C.H. Kim and T. Ozawa. New York: Plenum Press.

Esterbauer, H., M. Dieber-Rotheneder, G. Striegl, and G. Waeg. 1991. Role of vitamin E in preventing the oxidation of low-density lipoprotein. *Am. J. Clin. Nutr.* 53 (Suppl.): 314-321.

Esterbauer, H., G. Jürgens, O. Quehenberger, and E. Koller. 1987. Autoxidation of human low density lipoprotein: Loss of polyunsaturated fatty acids and vitamin E and generation of aldehydes. *J. Lipid Res.* 28: 495-509.

Esterbauer, H., H. Puhl, G. Waeg, A. Krebs, and M. Dieber-Rotheneder. 1993. The role of vitamin E in lipoprotein oxidation. In *Vitamin E in health and disease*, ed. L. Packer and J. Fuchs. New York: Marcel Dekker.

Esterbauer, H., G. Waeg, H. Puhl, and M. Dieber-Rotheneder. 1990. Mechanism of oxidation of low density lipoproteins. In *Drug metabolizing enzymes: Genetics, regulation and toxicology*, ed. M. Ingelman-Sundberg, J.-Å. Gustafsson, and S. Orrenius. Stockholm: Karolinska Institute.

Folkers, K., P. Langsjoen, R. Willis, P. Richardson, L.-J. Xia, C.-Q. Ye, and H. Tamagawa. 1990. Lovastatin decreases coenzyme Q levels in humans. *Proc. Natl. Acad. Sci. U.S.A.* 87: 8931-8934.

Food and Drug Administration. 1979. Vitamin and mineral drug products for over-the-counter human use. *Federal Register* 44: 152-231.

Forsberg, A., P. Tesch, A. Thorstensson, and J. Karlsson. 1976. Skeletal muscle fibers and athletic performance. In *Biomechanics*, ed. P.V. Komi. Baltimore: University Park Press.

Frankel, E.N., J. Kanner, J.B. German, E. Parks, and J.E. Kinsella. 1993. Inhibition of oxidation of human low-density lipoprotein by phenolic substances in red wine. *Lancet* 341: 454-457.

Franson, R.C., and M.D. Rosenthal. 1985. Oligomer of prostaglandin B_1 inhibit in vitro phospholipase A_2 activity. *Biochim. Biophys. Acta.* 1006: 272-277.

Fredholm, B.B., and J. Karlsson. 1970. Metabolic effects of prolonged sympathetic nerve stimulation in canine subcutaneous adipose tissue. *Acta Physiol. Scand.* 80: 567-576.

Frei, B., R. Stocker, and B.N. Ames. 1988. Antioxidant defenses and lipid peroxidation in human plasma. *Proc. Natl. Acad. Sci. U.S.A.* 85: 9748-9752.

Frei, B.B., and B.N. Ames. 1993. Relative importance of vitamin E in antiperoxidative defenses in human blood plasma and low-density lipoprotein (LDL). In *Vitamin E in health and disease*, ed. L. Packer and J. Fuchs. New York: Marcel Dekker.

Furuni, K., and N. Sugihara. 1994. Effect of cummene hydroperoxide on lipid peroxidation in cultured rat hepatocytes supplemented with eicosapentaenoic acid. *Biol. Pharm. Bull.* 17: 419-422.

Geigy. 1986. Constituents of living matters: Vitamins. *Geigy scientific tables 4*. Basel, Switzerland: CIBA-GEIGY Limited.

Geigy and Geigy. 1962. Vitamin C. In *Scientific tables*, ed. K. Diem, J.R. Basel, and S.A. Geigy.

Gey, K.F. 1993. Vitamin E and other essential antioxidants regarding coronary heart disease: Risk assessment studies. In *Vitamin E in health and disease*, ed. L. Packer and J. Fuchs. New York: Marcel Dekker.

Gey, K.F. 1986. On the antioxidant hypothesis with regard to arteriosclerosis. *Bibl. Nutr. Dieta* 37: 53-91.

Ghosh, C., R.M. Dick, and S.F. Ali. 1993. Iron/ascorbate-induced lipid peroxidation changes membrane fluidity and muscarinic cholinergic receptor binding in rat frontal cortex. *Neurochem. Int.* 23: 479-484.

Gohil, K., L. Packer, B. De Lumen, G.A. Brooks, and R.F. Burk. 1991. Vitamin E deficiency and vitamin C supplements: Exercise and mitochondrial oxidation. *J. Appl. Physiol.* 63: 1986-1991.

Goldspink, D.F. 1991. Exercise-related changes in protein turnover in mammalian striated muscle. *J. Exp. Biol.* 160: 127-148.

Goldstein, J.L., and M.S. Brown. 1987. Regulation of low density lipoprotein receptors: Implications for pathogenesis and therapy of hypercholesterolemia and atherosclerosis. *Circulation* 76: 504-507.

Gollnick, P.D., C.D. Ianuzzo, and D.W. King. 1971. Changes in muscle with exercise. In *Muscle metabolism during exercise*, ed. B. Pernow and B. Saltin, 69-85. New York: Plenum Press.

Gollnick, P.D., K. Piehl, C.W. Saubert, IV, and R.B. Armstrong. 1972. Diet, exercise and glycogen storage in human muscle fibers. *J. Appl. Physiol.* 33: 421-424.

Gomez-Fernandez, J.C., J. Villglain, and F.J. Aranda. 1989. Localization of alpha-tocopherol in membranes. *Ann. N.Y. Acad. Sci.* 570: 107-120.

Grandjean, A. 1994. Guidelines in dietary supplementation. *United States Olympic Committee's (USOC) Medical Commission. Consensus decision.* 1-4.

Green, R.L., S.S. Kaplan, B.S. Rabin, C.L. Stanitski, and U. Zdziarski. 1981. Immune function in marathon runners. *Ann. Allergy* 4: 73-75.

Greenberg, S., and W.H. Frishman. 1990. Co-enzyme Q_{10}: A new drug for cardiovascular disease. *J. Clin. Pharmacol.* 30: 596-608.

Grennhaff, P.L., A. Casey, A.H. Short, R. Harris, K. Söderlund, and E. Hultman. 1993. Influence of oral creatine supplementation of muscle torque during repeated bouts of maximal voluntary exercise in man. *Clin. Sci.* 84: 565-571.

Grootveld, M., E.B. Henderson, A. Farrell, and D.R. Blake. 1991. Oxidative damage to hyaluronate and glucose in synovial fluid during exercise of the inflamed rheumatoid joint: Detection of abnormal low-molecular-mass metabolites by proton-n.m.r. spectroscopy. *Biochem. J.* 273: 459-467.

Grune, T., W.G. Siems, K. Schonheit, and I.E. Blasig. 1993. Release of 4-hydroxynonenal, an aldehydic mediator of inflammation, during postischaemic reperfusion of the myocardium. *Int. J. Tissue React.* 15: 145-150.

Gryglewski, R., R. Palmer, and J. Moncada. 1986. Superoxide anion is involved in the breakdown of endothelium-derived vascular relaxing factor. *Nature* 320: 454-456.

Gudbjarnason, S., W. Braasch, C. Cowan, and R.J. Bing. 1968. Metabolism of infarcted heart muscle during tissue repair. *Am. J. Cardiol.* 22: 360-369.

Haglund, O., R. Luostarinen, R. Wallin, L. Wibell, and T. Saldeen. 1991. The effects of fish oil on triglycerides, cholesterol, fibrinogen and malondialdehyde in humans supplemented with vitamin E. *J. Nutr.* 121: 165-169.

Hallaq, H., T.W. Smith, and A. Leaf. 1992. Modulation of dihydropyridine-sensitive calcium channels in heart cells by fish oil fatty acids. *Proc. Natl. Acad. Sci. U.S.A.* 89: 1769-1774.

Halliwell, B. 1987. Oxidants and human disease: Some new concepts. *FASEB J.* 1: 358-364.

Halliwell, B., and M. Grootveld. 1987. The measurement of free radical reactions in humans: Some thoughts for future experimentation. *FEBS Lett.* 213: 9-14.

Halliwell, B., and J.M.C. Gutteridge. 1986. Oxygen free radicals and iron in relation to biology and medicine: Some problems and concepts. *Arch. Biochem. Biophys.* 246: 501-514.

———. 1989. *Free radicals in biology and medicine.* Oxford: Clarendon Press.

Halliwell, B., J.M.C. Gutteridge, and D.R. Blake. 1985. Metal ion and oxygen radical reactions in human inflammatory joint disease. *Philos. Trans. R. Soc. Lond.* 311: 659-671.

Harris, A., A.B. Robinson, and L. Pauling. 1973. Blood plasma L-ascorbic acid concentration for l-ascorbic acid dosages up to 12 grams a day. *Int. Res. Commun. Sys.* 1: 24-28.

Harris, R.C., K. Söderlund, and E. Hultman. 1992. Elevation of creatine in resting and exercised muscle of normal subjects by creatine supplementation. *Clin. Sci.* 83: 367-374.

Hatae, T., T. Yamada, T. Kobayashi, and I. Goto. 1991. Exercise-induced myalgia and high CKemia with a deletion in the dystrophin gene. *Rinsho Shinkeigaku* 31: 1155-1157.

Hatori, N., A. Miyazaki, H. Tadokoro, L. Rydén, R.E. Rajagopalan, M.C. Fishbein, S. Meerbaum, E. Corday, and J.K. Drury. 1989. Beneficial effects of coronary reperfusion, myocardial function, and infarct size in dogs. *J. Cardiovasc. Pharmacol.* 14: 396-404.

Hay, C.R.M., A.P. Durber, and R. Saynor. 1982. Effect of fish oil on platelet kinetics in patients with ischemic heart disease. *Lancet* 8284: 1269-1270.

Hellsten-Westing, Y., P.D. Balsom, B. Norman, and B. Sjödin. 1993. The effect of high-intensity training on purine metabolism in man. *Acta Physiol. Scand.* 149: 405-412.

Hellsten-Westing, Y., L. Kaijser, B. Ekblom, and B. Sjödin. 1994. Exchange of purines in human liver and skeletal muscle with short-time exhaustive exercise. *Am. J. Physiol.* 266: R81-R86.

Hellsten-Westing, Y., B. Norman, P.D. Balsom, and B. Sjödin. 1993. Decreased resting levels of adenine nucleotides in human skeletal muscle. *J. Appl. Physiol.* 74: 2523-2528.

Hellsten-Westing, Y., A. Sollevi, and B. Sjödin. 1991. Plasma accumulation of hypoxanthine, uric acid and creatine. *Eur. J. Appl. Physiol.* 62: 380-384.

Henriksen, O., J. Møgelvang, and C. Thomsen. 1988. Magnetic resonance in clinical physiology. *Clin. Physiol.* 8: 541-559.

Hertog, M.G.L., E.J.M. Feskens, P.C.H. Hollman, M.B. Katan, and D. Kromhout. 1993. Dietary antioxidant flavonoids and risk of coronary heart disease: The Zutphen Elderly Study. *Lipids* 342: 1007-1011.

Hillar, M., and A. Schwartz. 1972. Isolation and characterization of basic proteins (histones) from subcellular fractions of heart muscle, kidney and liver. *Acta Biochim. Pol.* 19: 277-285.

Hofman-Bang, C., N. Rehnqvist, and K. Swedberg. 1992. Coenzyme Q10 as an adjunctive treatment of congestive heart failure (for the Q10 Study Group 1992). *J. Am. Coll. Cardiol.* 19: 216A.

Holloszy, J. 1967. Biochemical adaptations in muscle: Effects of exercise on mitochondrial oxygen uptake and respiratory enzyme activity in skeletal muscle. *J. Biol. Chem.* 10: 2278-2282.

Holmgren, A., and P.-O. Åstrand. 1966. DL and the dimensions and functional capacities of the O_2-transport system in humans. *J. Appl. Physiol.* 21: 1463-1470.

Horrobin, D.F. 1986. Essential fatty acid and prostaglandin metabolism in Sjögren's syndrome, systemic sclerosis and rheumatoid arthritis. *Scand. J. Rheumatol.* 61: 242-245.

———. 1989. Effects of evening primrose oil in rheumatoid arthritis. *Ann. Rheum. Dis.* 48: 965-966.

Horwitt, M.K., C.C. Harvey, J. Dahm C.H. Jr., and M.T. Searcy. 1972. Relationship between tocopherol and serum lipid levels for determination of nutritional adequacy. *Ann. N.Y. Acad. Sci.* 203: 223-236.

Horwitt, M.K., W.H. Elliott, P. Kanjananggulpan, and C.D. Fitch. 1988. Serum concentrations of a-tocopherol after ingestion of various vitamin E preparations. *Am. J. Clin. Nutr.* 40: 240-245.

Hübner, C., G.F. Hoffmann, C. Charpenier, K.M. Gibson, B. Finckh, H. Puhk, H.-A. Lehr, and A. Kohlschütter. 1993. Decreased plasma ubiquinone-10 concentration in patients with mevalonate kinase deficiency. *Pediatr. Res.* 34: 129-133.

Huertas, J.R., M. Battino, G. Lenaz, and F.J. Mataix. 1991. Changes in mitochondrial and microsomal rat liver coenzyme Q9 and Q10 content induced by dietary fat and endogenous lipid peroxidation. *Feder. Proc. Eur. Biochem. Soc.* 287: 89-92.

Imagawa, M. 1990. Megavitamin therapy (coenzyme Q10 and vitamin B4) in Alzheimer's disease and senile dementia of Alzheimer type. *Adv. Behav. Biol.* 38B: 489-491.

Inoe, T., Z. Mu, K. Sumikawa, K. Adachi, and T. Okochi. 1993. Effect of physical exercise on the content of 8-hydroxydeoxyguanosine in nuclear DNA prepared from lymphocytes. *Jpn. J. Cancer Res.* 84: 720-725.

International Olympic Committee, and Medical Commission. 1976. *Medical Commission booklet.* Montreal: The International Olympic Committee.

Jacobs, I., N. Westlin, J. Karlsson, M. Rasmusson, and B. Houghton. 1982. Muscle glycogen and diet in elite soccer players. *Eur. J. Appl. Physiol.* 48: 297-302.

Janssen, G.M.E., H.R. Scholte, M.H.M. Vandrager-Verduin, and R.J.M.M. Does. 1989. Muscle carnitine level in endurance training and running a marathon. *Int. J. Sports Med.* 10 (Suppl. 3): S153-S155.

Janssen, G.M.E., J.W.J. Wersch, V. Kaiser, and R.J.M.M. Does. 1989. White cell system changes associated with a training period of 18-20 months: A transverse and a longitudinal approach. *Int. J. Sports Med.* 10 (Suppl. 3): S176-S180.

Jenkins, R. 1986. Potentially harmful effect of ambient oxygen [Abstract]. *Med. Sci. Sports Exerc.* 18: S67.

Jenkins, R.R. 1993. Exercise, oxidative stress and antioxidants: A review. *Int. J. Sport Nutr.* 3: 356-375.

Jenkins, R.R., R. Friedland, and H. Howald. 1984. The relationship of oxygen uptake to superoxide dismutase and catalase activity in human skeletal muscle. *Int. J. Sports Med.* 5: 11-14.

Jenkins, R.R., and A. Goldfarb. 1993. Introduction: Oxidant stress, aging and exercise. *Med. Sci. Sports Exerc.* 25: 210-212.

Jenkins, R.R., K. Krause, and L.S. Schofield. 1993. Influence of exercise on clearance of oxidant stress products and loosely bound iron. *Med. Sci. Sports Exerc.* 25: 213-217.

Jenkins, R.R.H. and B. Halliwell. 1994. Metal binding agents: Possible role in exercise. In *Exercise and oxygen toxicity*, ed. C.K. Sen, L. Packer, and O. Hännie. Amsterdam: Elsevier.

Ji, L.L. 1993. Antioxidant enzyme responses to exercise and aging. *Med. Sci. Sports Exerc.* 25: 225-231.

Jöbsis, F.F. 1964. Basic process in cellular respiration. In *Handbook of physiology*, ed. W.O. Fenn and H. Rahn. Washington, DC: American Physiology Society.

Johansen, K., H. Theorell, J. Karlsson, B. Diamant, and K. Folkers. 1991. Coenzyme Q_{10}, alpha-tocopherol and free cholesterol in HDL and LDL fractions. *Ann. Med.* 23: 649-656.

Johansson, C. 1987. Elite sprinters, ice hockey players, orienteers and marathon runners. MD diss., Umeå University, Umeå, Sweden.

Jones, B.H., D.N. Cowan, and J.J. Knapik. 1994. Exercise, training and injuries. *Sports Med.* 18: 202-214.

Kalén, A., E.-L. Appelkvist, and G. Dallner. 1989. Age-related changes in the lipid composition of rat and human tissues. *Lipids* 24: 579-584.

Kalén, A., B. Norling, E.-L. Appelkvist, and G. Dallner. 1987. Ubiquinone synthesis in the microsomal fraction of rat liver. *Biochim. Biophys. Acta* 926: 70-80.

Kamei, M., T. Fujita, T. Kanbe, K. Sasaki, K. Oshiba, S. Otani, I. Matsu-Yuasa, and S. Morisawa. 1986. The distribution and content of ubiquinone in foods. *Int. J. Vitam. Nutr. Res.* 36: 57-63.

Kannel, W.B. 1988. Contributions of the Framingham Study to the conquest of coronary artery disease. *Am. J. Cardiol.* 62: 1109-1112.

Kanner, J., S. Harrel, and B. Hazan. 1986. Muscle membranal lipid peroxidation by an iron redox cycle system: Initiation by oxyradicals and site-specific mechanisms. *J. Agric. Food Chem.* 34: 506-510.

Kanter, M.M. 1994. Free radicals, exercise and antioxidant supplementation. *Int. J. Sport Nutr.* 4: 205-220.

Kanter, M.M., G.R. Lesmes, L.A. Kaminsky, and J. LaHam-Salger. 1988. Serum creatine kinase and lactate dehydrogenase changes following an eighty kilometer race. *Eur. J. Appl. Physiol.* 57: 60-83.

Kanter, M.M., L.A. Nolte, and J.O. Holloszy. 1993. Effects of antioxidant vitamin mixture on lipid peroxidation at rest and postexercise. *J. Appl. Physiol.* 74: 965-969.

Karlson, P. 1965. *Introduction to modern biochemistry.* New York: Academic Press.

Karlsson, J. 1971. Lactate and phosphagen concentrations in working muscle of man. *Acta Physiol. Scand. Suppl.* 358: 1-72.

———. 1979. Localized muscular fatigue: Role of muscle metabolism and substrate depletion. In *Exercise and sports sciences review*, ed. R.S. Hutton and D.I. Miller. Philadelphia: Franklin Institute Press.

———. 1986a. Muscle exercise, energy metabolism and blood lactate. In *The anaerobic threshold: Physiological and clinical significance*, ed. L. Tavazzi and P.E. di Prampero. Basel, Switzerland: Karger.

———. 1986b. Muscle fiber composition, metabolic potentials, oxygen transport and exercise performance in man. In *Biochemical aspects of physical exercise*, ed. G. Benzi, L. Packer, and N. Siliprandi. Amsterdam: Elsevier.

———. 1987. Heart and skeletal muscle ubiquinone or CoQ_{10} as a protective agent against radical formation in man. *Advances in myochemistry.* London: John Libbey.

———. 1993. *Vitamin Q and our health.* Vaxholm, Sweden: OBLA AB.

———. 1995. *Längdskidåkning och kostförstärkning. Idrott och kost.* Vaxholm, Sweden: OBLA AB.

———. In press. Plasma nutrients in elite cross-country skiers after nutratherapy. *Int. J. Sports Nutr.*

Karlsson, J., S. Branth, and J. Ekstrand. 1994. *Elitfotboll och kostförstärkning; utredning på VM-laget 1994. Idrott och kost.* Vaxholm, Sweden: OBLA AB.

Karlsson, J., and B. Diamant. 1992. Kostförstärkning - vetenskap eller myt? *Svensk Skidsport* 23: 64-67.

Karlsson, J., B. Diamant, P.-O. Edlund, B. Lund, K. Folkers, and H. Theorell. 1992. Plasma ubiquinone, alpha-tocopherol and cholesterol in man (in Swedish). *Int. J. Vitam. Nutr. Res.* 62: 160-164.

Karlsson, J., B. Diamant, K. Folkers, P.-O. Edlund, B. Lund, and H. Theorell. 1990a. Plasma ubiquinone and cholesterol contents with and without ubiquinone treatment. In *Highlights in ubiquinone research*, ed. G. Lenaz, O. Barnabei, and M. Battino. London: Taylor and Francis.

———. 1990b. Skeletal muscle and blood CoQ_{10} in health and disease. *Highlights in ubiquinone research*. London: Taylor and Francis.

Karlsson, J., B. Diamant, K. Folkers, and B. Lund. 1991. Muscle fiber types, ubiquinone content and exercise capacity in hypertension and effort angina. *Ann. Med.* 23: 339-344.

Karlsson, J., B. Diamant, H. Theorell, and K. Folkers. 1991. Skeletal muscle coenzyme Q_{10} in healthy man and selected patient groups. In *Biomedical and clinical aspects of coenzyme Q*, ed. K. Folkers, G.P. Littarru, and T. Yamagami. Amsterdam: Elsevier.

———. 1993. Ubiquinone and α-tocopherol in plasma; means of translocation or depot. *Clin. Invest.* 71: S84-S91.

Karlsson, J., B. Diamant, H. Theorell, K. Johansen, and K. Folkers. 1993. Plasma alpha-tocopherol and ubiquinone and their relations to muscle function in healthy humans and in cardiac diseases. In *Vitamin E: Biochemistry and clinical applications*, ed. L. Packer and J. Fuchs. New York: Marcel Dekker.

Karlsson, J., K. Frith, B. Sjödin, P.D. Gollnick, and B. Saltin. 1974. Distribution of LDH isozymes in human skeletal muscle. *Scand. J. Clin. Lab. Invest.* 33: 307-312.

Karlsson, J., L. Lin, S. Gunnes, C. Sylvén, and H. Åström. 1996. Muscle ubquinone in male effort angina patients. *Mol. Cell. Biochem.* 156: 173-178.

Karlsson, J., L. Lin, C. Sylvén, and E. Jansson. 1996. Muscle ubquinone in healthy physically active males. *Mol. Cell. Biochem.* 156: 169-172.

Karlsson, J., P. Lobstein, G.H. Templeton, and J.T. Willerson. 1974. Lactate metabolism and acute canine myocardial ischemia. *Recent Adv. Stud. Cardiac Struct. Metab.* 3: 713-720.

Karlsson, J., L.-O. Nordesjö, and B. Saltin. 1974. Muscle glycogen utilization during exercise after physical training. *Acta Physiol. Scand.* 90: 210-217.

Karlsson, J., H. Rønneberg, and R. Rønneberg. 1996. Endurance athletes, insulin secretion-sensitivity and muscle quality. *Can. J. Cardiol.* In press.

Karlsson, J., M. Rasmusson, A. von Schevelow, P. Nilsson-Ehle, and B. Diamant. 1992. *Utförsåkarens kost. Idrott och kost*. Vaxholm, Sweden: OBLA AB.

Karlsson, J., S. Rosell, and B. Saltin. 1972. Carbohydrate and fat metabolism in contracting canine skeletal muscle. *Pflügers Arch.* 331: 57-69.

Karlsson, J., and B. Saltin. 1971. Diet, muscle glycogen, and endurance performance. *J. Appl. Physiol.* 31: 203-206.

Karlsson, J. and B. Semb. 1996. Muscle fibers, ubiquinone and exercise capacity in effort angina. *Mol. Cell. Biochem.* 156:179-184.

Karlsson, J., B. Sjödin, A. Thorstensson, B. Hultén, and K. Frith. 1975. LDH isozymes in skeletal muscle of endurance and strength trained subjects. *Acta Physiol. Scand.* 93: 150-156.

Karlsson, J., and H.J. Smith. 1984. Muscle fibers in human skeletal muscle and their metabolic and circulatory significance. In *The peripheral circulation*, ed. S. Hunyor, J. Ludbrook, J. Shaw, and H. McGrath. Amsterdam: Excerpta Medica.

Karlsson, J., G.H. Templeton, and J.T. Willerson. 1973. Relationship between epicardial S-T segment changes and myocardial metabolism during acute coronary insufficiency. *Circ. Res.* 32: 725-730.

Keys, A., C. Aravanis, and H. Blackburn. 1967. Epidemiological studies related to coronary heart disease: Characteristics of men aged 40-59 in seven countries. *Acta Med. Scand.* (Suppl.): 1-392.

Kiessling, K.-H., L. Pilström, J. Karlsson, and K. Piehl. 1973. Mitochondrial volume in skeletal muscle from young and old physically untrained and trained healthy men and from alcoholics. *Clin. Sci.* 44: 547-554.

Kim, H., X. Chen, and C.N. Gillis. 1992. Ginsenosides protect pulmonary endothelium against free radical-induced injury. *Biochem. Biophys. Res. Commun.* 189: 670-676.

King, T.E. 1990. Preparation, properties and reconstitution of QP-C. In *Highlights in ubiquinone research*, ed. G. Lenaz, O. Barnabei, A. Rabbi, and M. Battino. London: Taylor and Francis.

King, T.E., Y. Xu, T.Y. Wang, and W.H. Ding. 1986. A mitochondrial coenzyme Q protein - QP-C. In *Biomedical and clinical aspects of coenzyme Q*, ed. K. Folkers and Y. Yamamura. Amsterdam: Elsevier Science.

Komi, P.V. 1987. Neuromuscular performance: Considerations for basic mechanisms and adaptive responses. *Int. J. Sports Med.* 8 (Suppl. 1): 1-70.

Komi, P.V., and J. Karlsson. 1979. Physical performance, skeletal muscle enzyme activities and fibre types in monozygous and dizygous twins of both sexes. *Acta Physiol. Scand. Suppl.* 462: 1-28.

Kowdley, K.V., J.B. Mason, S.N. Meydani, S. Cornwall, and R.J. Grand. 1992. Vitamin E deficiency and impaired cellular immunity related to intestinal fat malabsorption. *Gastroenterology* 102: 2139-2142.

Kühnau, J. 1976. The flavonoids. A class of semi-essential food components: Their role in human nutrition. *World Rev. Nutr. Diet.* 24: 117-120.

Kuipers, H., G.M.E. Janssen, F. Bosman, P.M. Frederick, and P. Geurten. 1989. Structural and ultrastructural changes in skeletal muscle associated with long-distance training and running. *Int. J. Sports Med.* 10 (Suppl. 3): S156-S159.

Leaf, A., and P.C. Weber. 1988. Cardiovascular effects of n-3 fatty acids. *N. Engl. J. Med.* 318: 549-557.

Lehninger, A.L. 1965. *The mitochondrion.* New York: W.A. Benjamin.

Lester, R.L., and F.L. Crane. 1959. The natural occurrence of coenzyme Q and related compounds. *J. Biol. Chem.* 234: 2169-2175.

Leventhal, L.J., E.G. Boyce, and R.B. Zurier. 1993. Treatment of rheumatoid arthritis with gammalinolenic acid. *Ann. Intern. Med.* 119: 867-873.

Levine, R.L., D. Garland, C.N. Oliver, A. Amici, I. Climent, A.-G. Lenz, B.-W. Ahn, S. Shaltiel, and E.R. Stadtman. 1990. Determination of carbinyl content in oxidatively modified protein. *Methods Enzymol.* 186: 464-478.

Lin, L., P. Sotonyi, E. Somogyi, J. Karlsson, K. Folkers, Y. Nara, C. Sylvén, L. Kaijser, and E. Jansson. 1988. Coenzyme Q10 content in different parts of the normal human heart. *Clin. Physiol.* 8: 391-398.

Lisook, A.B. 1990. FDA audits of clinical studies and procedure. *J. Clin. Pharmacol.* 30: 296-302.

Lithell, H., M. Cedermark, J. Froberg, P. Tesch, and J. Karlsson. 1981. Increase in lipoprotein lipase activity in skeletal muscle during heavy exercise: Relation to epinephrine excretion. *Metabolism* 30: 1130-1134.

Lithell, H., J. Örlander, R. Schéle, B. Sjödin, and J. Karlsson. 1979. Changes in lipoprotein-lipase activity and lipid stores in human skeletal muscle with prolonged heavy exercise. *Acta Physiol. Scand.* 107: 257-261.

Littarru, G.P., L. Ho, and K. Folkers. 1972. Deficiency of coenzyme Q10 in human heart disease. *Int. J. Vitam. Nutr. Res.* 42: 291-297.

Lopez, L. 1990. Photoinduced electron transfer. *Top. Curr. Chem.* 156: 117-166.

Luft, R. 1994. The development of mitochondrial medicine. *Proc. Natl. Acad. Sci. U.S.A.* 91: 8731-8738.

Luft, R., D. Ikkos, G. Palmieri, L. Ernster, and B. Afzelius. 1962. A case of severe hypermetabolism of nonthyroid origin with a defect in the maintenance of mitochondrial respiratory control: A correlated clinical, biochemical and morphological study. *J. Clin. Invest.* 41: 1776-1801.

Maguire, J.J., V. Kagan, B.A. Ackrell, E. Serbinova, and L. Packer. 1992. Succinate-ubiquinone reductase linked recycling of alpha-tocopherol in reconstituted systems and mitochondria: Requirement for reduced ubiquinone. *Arch. Biochem. Biophys.* 292: 47-53.

Martinez, M.C., F. Bosch-Morell, A. Raya, and J. Rima. 1994. 4-Hydroxynonenal, a lipid peroxidation product, induces relaxation of human cerebral arteries. *J. Cereb. Blood Flow Metab.* 14: 693-696.

McCord, J.M. 1974. Free radicals and inflammation protection of synovial fluid by SOD. *Science* 185: 529-531.

———. 1985. Oxygen-derived free radicals in post-ischemic tissue injury. *N. Engl. J. Med.* 312: 159-163.

McMurchie, E.J., and G.H. McIntosh. 1986. Thermotropic interaction of vitamin E with dimyristoyl and dipalmitoyl phosphatidylcholine liposomes. *J. Nutr. Sci. Vitaminol.* 32: 551-558.

Mead, J.C., and J. Mertin. 1988. Fatty acids and immunity. *Adv. Lipid Res.* 21: 103-108.

Mellander, S. 1981. Differentiation of fiber composition, circulation, and metabolism in limb muscles of dog, cat and man. In *Vasodilation*, ed. P.M. Vanhuette and I. Leusen. New York: Raven Press.

Merry, P., M. Grootveld, J. Lunec, and D. Blake. 1991. Oxidative damage to lipids within the inflamed human joint provides evidence of radical-mediated hypoxic-reperfusion injury. *Am. J. Clin. Nutr.* 53: 362S-369S.

Meydani, M., W.J. Evans, G. Handelman, and L. Biddle. 1993. Protective effect of vitamin E on exercise-induced oxidative damage in young and older adults. *Am. J. Physiol.* 264: 992-998.

Meydani, S.N., M.P. Barklund, and S. Liu. 1989. Effect of vitamin E supplementation on immune responsiveness of healthy elderly subjects. *Ann. N.Y. Acad. Sci.* 570: 283-290.

———. 1990. Vitamin E supplementation enhances cell-mediated immunity in healthy elderly subjects. *Am. J. Clin. Nutr.* 52: 557-563.

Meydani, S.N., M. Hayek, and L. Coleman. 1992. Influence of vitamin E and B_6 on immune response. *Ann. N.Y. Acad. Sci.* 669: 125-139.

Michaelis, L. 1946. *Currents in biochemical research.* New York: Interscience.

Mitchell, J.H. 1985. Cardiovascular control during exercise: Central and reflex neural mechanisms. *Am. J. Cardiol.* 55 (Suppl. D): 34-41.

Mitchell, J.H., and R.F. Schmidt. 1983. Cardiovascular reflex control by afferent fibers from skeletal muscle receptors. In *Handbook of Physiology - the Cardiovascular System* Vol. 3, ed. J.T. Shepherd and F.M. Abboud. S. Bethesda, MD: American Physiological Society.

Mitchell, J.H., W.C. Reardon, D. McCloskley, and K. Wildenthal. 1977. Possible role of muscle receptors in the cardiovascular response to exercise. In *The marathon: Physiological, medical, epidemiological and psychological studies*, ed. P. Milvy. New York: New York Academy of Sciences.

Mitchell, P. 1976. Possible molecular mechanisms of the protonmotive function of cytochrom systems. *J. Theor. Biol.* 62: 327-367.

————. 1991. The vital protonmotive role of coenzyme Q. In *Biomedical and clinical aspects of coenzyme Q*, ed. K. Folkers, G.P. Littarru, and T. Yamagami. Amsterdam: Elsevier Science.

Moll, W., and H. Bartels. 1968. The diffusion coefficient of myoglobin in muscle homogenates. *Pflügers Arch.* 299: 247-254.

Morse, P.F., D.F. Horrobin, M.S. Manku, J.C.M. Stewart, R. Allen, S. Littlewood, S. Wright, J. Burton, D.J. Gould, P.J. Holt, C.T. Jansen, L. Mattila, W. Meigel, T.H. Dettke, D. Wexler, L. Guenther, A. Bordoni, and A. Patrizi. 1989. Meta-analysis of placebo-controlled studies of the treatment of atopic eczema. Relationship between plasma essential fatty acid changes and clinical response. *Br. J. Dermatol.* 121: 75-90.

Mortensen, S.-A. 1993. Perspectives on therapy of cardiovascular artery diseases with coenzyme Q_{10} (ubiquinone). *Clin. Invest.* 71: S116-S123.

Mortensen, S.-A., P. Heidt, and J. Sehested. 1990. Clinical perspectives in the treatment of cardiovascular diseases with coenzyme Q_{10}. In *Highlights in ubiquinone research*, ed. G. Lenaz, O. Barnabei, A. Rabbi, and M. Battino. London: Taylor and Francis.

Mortensen, S.-A., S. Vadhanavikit, U. Baandrup, and K. Folkers. 1985. Long-term coenzyme Q_{10} therapy: A major advance in the management of resistant myocardial failure. *Drugs Exp. Clin. Res.* 11: 581-593.

Morton, R.A., G.M. Wilson, J.S. Lowe, and W.M.F. Leat. 1957. Ubiquinone. *Chemical Industry* 1649.

Mugge, A., J.H. Elwell, T.E. Peterson, and D.G. Harrison. 1991. Release of intact endothelium-derived relaxing factor depends on endothelium superoxide dismutase activity. *Am. J. Physiol.* 260: 19-25.

Muller, D., L. Harris, and J. Lloyd. 1974. The relative importance of the factors involved in the absorption of vitamin E in children. *Gut* 15: 966-971.

Murphy, S.P., A.F. Subar, and G. Block. 1990. Vitamin E intakes and sources in the United States. *Am. J. Clin. Nutr.* 52: 361-367.

Ng, C.K., C.J. Handley, B.N. Preston, and H.C. Robinson. 1992. The extracellular processing and catabolism of hyaluronan in cultured adult articular cartilage explants. *Arch. Biochem. Biophys.* 298: 70-79.

Nikkila, E.A., T. Kuusi, and M.-R. Taskinen. 1982. Role of lipoproteinlipase and hepatic endothelial lipase in the metabolism of high density lipoproteins: A novel concept on cholesterol transport in HDL cycle. In *Metabolic risk factors in ischemic cardiovascular disease*, ed. L.A. Carlsson and B. Pernow. New York: Raven Press.

Nilsson-Ehle, P., A.S. Garfinkel, and M.C. Scholtz. 1980. Lipolytic enzymes and plasma lipoprotein metabolism. *Ann. Rev. Biochem.* 49: 667-693.

NLN (Nordisk Läkemedelsämnden). 1993. Good clinical trial practice. Nordic guidelines. *NLN Publications,* 28:1-36.

Noberasco, G., P. Odetti, D. Boeri, M. Maiello, and L. Adezatti. 1991. Malondialdehyde (MDA) level in diabetic subjects. *Biomed. Pharmacother.* 45: 193-196.

Nohl, H. 1986. Oxygen free radical release in mitochondria: Influence of age. In *Free radicals, aging and degenerative diseases,* ed. J.E. Johnson, R. Walford, D. Harman, and J. Miquel. New York: Liss.

NRC (National Research Council). 1989a. *Recommended dietary allowances.* Washington, DC: National Academy of Sciences.

NRC (National Research Council). 1989b. *Diet and health: Implications for reducing chronic disease risk.* Washington, DC: National Academy Press.

Okamoto, T., T. Matruya, Y. Fukunaga, T. Kishio, and T. Yamagami. 1989. Human serum ubiquinol-10 levels and relationship to serum lipids. *Int. J. Vitam. Nutr. Res.* 59: 288-292.

Oliwiecki, S., J.L. Burton, K. Elles, and D.F. Horrobin. 1990. Levels of essential and other fatty acids in plasma and red cell phospholipids from normal controls and patients with atopic eczema. *Acta Derma. Venereol.* 71: 224-228.

Opie, L.H. 1965. Effect of extracellular pH on function and metabolism of isolated perfused rat hearts. *Am. J. Physiol.* 209: 1075-1080.

Örlander, J., K.-H. Kiessling, J. Karlsson, and B. Ekblom. 1977. Low intensity training, inactivity and resumed training in sedentary men. *Acta Physiol. Scand.* 101: 351-362.

Örlander, J., K.-H. Kiessling, L. Larsson, J. Karlsson, and A. Aniansson. 1977. Skeletal muscle metabolism and ultrastructure in relation to age in sedentary men. *Acta Physiol. Scand.* 104: 249-261.

Packer, L. 1986. Oxygen radicals and antioxidants in endurance exercise. In *Biochemical aspects of physical exercise,* ed. G. Benzi, L. Packer, and N. Siliprandi. Amsterdam: Elsevier.

Packer, L., and J. Fuchs, eds. 1993. *Vitamin E in health and disease.* New York: Marcel Dekker.

Packer, L., and C. Viguie. 1989. Human exercise: Oxidative stress and antioxidant therapy. In *Advances in myochemistry.* Vol. 2, ed. G. Benzi. London: John Libbey.

Perly, B.I., C.P. Smith, L. Hughes, G.W. Burton, and K.U. Ingold. 1985. Estimation of the location of natural alpha-tocopherol in lipid bilayers by C13-NMR spectroscopy. *Biochim. Biophys. Acta* 819: 131-135.

Piehl-Aulin, K., C. Laurent, A. Engström-Laurent, S. Hellström, and J. Henriksson. 1991. Hyaluronan in human skeletal muscle of lower extremity: Concentration, distribution and effect of exercise. *J. Appl. Physiol.* 71: 2493-2498.

Pincemail, J., C. Deby, and A. Dethier. 1987. Pentane measurements in man as an index of lipoperoxidation. *Bioelectrochem. Bioenerg.* 18: 117-125.

Pruzanski, W., P. Vadas, E. Stefanski, and M.B. Urowitz. 1985. Phospholipase A_2 activity in sera and synovial fluids in rheumatoid arthritis: Its possible role as a proinflammatory enzyme. *J. Rheumatol.* 12: 211-216.

Pryor, W.A. 1993. The role of vitamin E in the protection of in vitro systems in animals against the effects of ozone. In *Vitamin E: Biochemistry and clinical applications*, ed. L. Packer and J. Fuchs. New York: Marcel Dekker.

Ramasarma, T. 1985. Natural occurrence and distribution of coenzyme Q. In *Coenzyme Q*, ed. G. Lenaz. Chichester, England: John Wiley & Sons.

Rånby, B., and J.F. Rabek. 1978. *Singlet oxygen: Reactions with organic compounds and polymers*. Chichester, England: John Wiley & Sons.

Ravnskov, U. 1992. Cholesterol lowering trials in coronary heart disease: Frequency of citation and outcome. *Brit. Med. J.* 305: 15-18.

Riely, C.A., and G. Cohen. 1974. Pentane measurement in man as an index of lipoperoxidation. *Science* 183: 208-210.

Rimm, E.B., A. Ascherio, W.C. Willett, E.L. Giovannucci, and M.J. Stampfer. 1992. Vitamin E supplementation and risk of coronary heart disease among men. *Circulation* 86 (Suppl. I-463): Abstr. No. 1848.

Rocklin, R.E., L. Thistle, L. Gallant, M.S. Manku, and D. Horrobin. 1986. Altered arachidonic acid content in polymorphonuclear and mononuclear cells from patients with allergic rhinitis and/or asthma. *Lipids* 21: 17-20.

Rokitzki, L., E. Logemann, G. Huber, E. Keck, and J. Keul. 1994. α-tocopherol supplementation in racing cyclists. *Int. J. Sport Nutr.* 4: 253-264.

Rowley, D.A., J.M.C. Gutteridge, D.R. Blake, M. Farr, and B. Halliwell. 1984. Lipid peroxidation in rheumatoid arthritis: Thiobarbituric acid reactive material and catalytic iron salts in synovial fluid from rheumatoid patients. *Clin. Sci.* 66: 691-695.

Sahlin, K., K. Ekberg, and S. Cizinsky. 1991. Changes in plasma hypoxanthine and free radical markers during exercise in man. *Acta Physiol. Scand.* 142: 275-281.

Salonen, J.T., R. Salonen, and K. Seppänen. 1988. Relationship of serum selenium and antioxidants to plasma lipoproteins, platelet aggregability and prevalent ischaemic heart disease in eastern Finnish men. *Atherosclerosis* 70: 155-160.

Saltin, B. 1973. Metabolic fundamentals in exercise. *Med. Sci. Sports* 5: 137-146.

———. 1985. Hemodynamic adaptations to exercise. *Am. J. Cardiol.* 55 (Suppl. D): 42-47.

Saltin, B., G. Blomqvist, J.H. Mitchell, R.L. Johnson, K. Wildenthal, and C.B. Chapman. 1968. Response to exercise after bed rest and training. *Circulation* 37-38 (Suppl. VII): 1-78.

Savard, G., B. Kiens, and B. Saltin. 1987. Central cardiovascular factors as limits to endurance; with a note on the distinction between maximal oxygen uptake and endurance fitness. *Exercise: Benefits, limits and adaptations*. Edinburgh: E. & F.N. Spon.

Sawyer, D.T. 1988. O_2! Who would have imagined all the biological processes that involve oxygen? *CHEMTECH* (June): 369-375.

Schrader, J., S. Nees, and E. Gerlach. 1977. Evidence for a cell surface adenosine receptor on coronary myocytes and atrial muscle cell. *Pflügers Arch.* 369: 251-257.

Scott, G. 1995. Antioxidants, the modern elixir? *Chemistry in Britain.* (November): 879-882.

Serbinova, E.A., M. Tsuchiya, S. Goth, V.E. Kagan, and L. Packer. 1993. Antioxidant action of α-tocopherol and α-tocotrienol in membranes. In *Vitamin E in health and disease*, ed. L. Packer and J. Fuchs. New York: Marcel Dekker.

Shepherd, J.T., C.G. Blomqvist, A.R. Lind, J.H. Mitchell, and B. Saltin. 1981. Static (isometric) exercise: Retrospection and introspection. *Circ. Res.* 48 (Suppl. I): 179-188.

Shigenaga, M.K., C.J. Gimeno, and B.N. Ames. 1989. Urinary 8-hydroxy-2'-deoxyguanosine as a biological marker of *in vivo* oxidative damage. *Proc. Natl. Acad. Sci. U.S.A.* 86: 9697-9701.

Shimokawa, H., and P.M. Vanhoutte. 1988. Dietary cod-liver oil improves endothelium dependent responses in hypercholesterolemic and atherosclerotic porcine coronary arteries. *Circulation* 78: 1421-1430.

Sies, H. 1993. Strategies of the antioxidant defence. *Eur. J. Biochem.* 215: 213-219.

Sies, H., S. Kaiser, P. Di Mascio, and M.E. Murphy. 1992. Scavenging of singlet molecular oxygen by tocopherols. In *Vitamin E: Biochemistry and clinical applications*, ed. L. Packer and J. Fuchs. New York: Marcel Dekker.

Simon-Schnass, I.M., and L. Korniszewski. 1990. The influence of vitamin E on rheological parameters in high altitude mountaineers. *Int. J. Vitam. Nutr. Res.* 60: 26-34.

Simon-Schnass, I.M., and H. Pabst. 1988. Influence of vitamin E on physical performance. *Int. J. Vitam. Nutr. Res.* 58: 49-54.

Simon-Schnass, I.M., J. Reiman, and V. Böhlau. 1984. Vitamin-E-therapie. *Notabene Medici* 14: 793-794.

Sjödin, B. 1976. Lactate dehydrogenase in human skeletal muscle. *Acta Physiol. Scand. Suppl.* 436: 1-32.

Sjöström, M., C. Johansson, and R. Lorentzon. 1987. Muscle pathomorphology in m. quadriceps of marathon runners: Early signs of strain disease or functional adaptation? *Acta Physiol. Scand.* 132: 537-541.

Skjervold, H. 1991. Lifestyle diseases - human diet. *Meieriosten* 19: 527-529.

Sollevi, A. 1986. Cardiovascular effects of adenosine in man: Possible clinical applications. *Prog. Neurobiol.* 27: 319-349.

Søyland, E. 1993. The effect of very long-chain n-3 fatty acids on immunorelated skin diseases and some immune reactions. PhD diss., Oslo University, Norway.

Søyland, E., M.S. Nenseter, L.R. Braathen, and C.A. Drevon. 1993. Very long-chain n-3 and n-6 polyunsaturated fatty acids inhibit proliferation of human T-lymphocytes in vitro. *Eur. J. Clin. Invest.* 23: 112-121.

Stadtman, E.R., and C.N. Oliver. 1991. Metal-catalyzed oxidations of proteins: Physiological consequences. *J. Biol. Chem.* 266: 3341-3346.

Stampfer, M.J., J.-A.E. Manson, G.A. Colditz, F.E. Speizer, W.C. Willett, and C.H. Hennekens. 1992. A prospective study on vitamin E supplementation and risk of coronary disease in women. *Circulation* 86 (Suppl. I-463): Abstr. No. 1847.

Stocker, R., V.W. Bowry, and B. Frei. 1991. Ubiquinol-10 protects human low density lipoprotein more efficiently against lipid peroxidation than does alpha-tocopherol. *Proc. Natl. Acad. Sci. U.S.A.* 88: 1646-1650.

Stocker, R., and B. Frei. 1991. Endogenous antioxidant defenses in human blood plasma. In *Oxidative stress: Oxidants and antioxidants*, ed. H. Sies. Orlando, FL: Academic Press.

Stubbe, I.G., A. Gustafsson, and P. Nilsson-Ehle. 1982. Alterations in plasma proteins and lipoproteins in acute myocardial infarction: Effects on activation of lipoprotein lipase. *Scand. J. Clin. Lab. Invest.* 42: 437-444.

Sumida, S., K. Tanaka, H. Kiato, and F. Nakadomo. 1989. Exercise induced lipid peroxidation and leakage of enzyme before and after vitamin E supplementation. *Int. J. Biochem.* 21: 835-838.

Tappel, A. 1980. Vitamin E and selenium from *in vivo* lipid peroxidation. *Ann. N.Y. Acad. Sci.* 355: 18-31.

Taylor, P.R., S. Dawsey, and D. Albanes. 1990. Cancer prevention trials in China and Finland. *Ann. Epidemiol.* 1: 195-203.

Tengerdy, R.P. 1989. The effect of vitamin E on immune response and disease resistance. *Ann. N.Y. Acad. Sci.* 570: 335-344.

Thompson, J.A., K.D. Anderson, J.M. Dipietro, J.A. Zwiebel, M. Zametta, W.F. Anderson, and T. Maciag. 1988. Site-directed neovessel formation *in vivo*. *Science* 241: 1349-1352.

Tiidus, P.M., and M.E. Houston. 1993. Vitamin E status does not affect the responses to exercise training and acute exercise in female rats. *J. Nutr.* 123: 834-840.

———. 1994. Antioxidant and oxidative enzyme adaptations to vitamin E deprivation and training. *Med. Sci. Sports Exerc.* 26: 354-359.

Traber, M.G., W. Cohn, and D.P.R. Muller. 1993. Absorption, transport and delivery to tissues. In *Vitamin E in health and disease*, ed. L. Packer and J. Fuchs. New York: Marcel Dekker.

Traber, M.G., and H.J. Kayden. 1984. Vitamin E is delivered to cells via the high affinity receptor for low density lipoprotein. *Am. J. Nutr.* 40: 747-751.

Van der Beck, E.J. 1985. Vitamins and endurance training: Food for running and faddish claims. *Sports Med.* 2: 175-181.

van der Vusse, G.J., G.M.E. Janssen, W.A. Coumans, H. Kuipers, R.J.M.M. Does, and F. ten Hoor. 1989. Effects of training and 25-, 25- and 42-km contests on the skeletal muscle content of adenine and guanine nucleotides, creatine phosphate, and glycogen. *Int. J. Sports Med.* 10 (Suppl. 3): S146-S152.

van Erp-Baart, A.M.J., W.M.H. Saris, R.A. Binkhorst, J.A. Vos, and J.W.H. Elvers. 1989a. Nationwide survey on nutritional habits in elite athletes. Part I. Energy, carbohydrate, protein and fat intake. *Int. J. Sports Med.* 10 (Suppl. 1): S3-S10.

———. 1989b. Nationwide survey on nutritional habits in elite athletes. Part II. Mineral and vitamin intake. *Int. J. Sports Med.* 10 (Suppl. 1): S11-S16.

Vanhoutte, P.M. 1991. Hypercholesterolaemia, atherosclerosis and release of endothelium-derived relaxing factor by aggregating platelets. *Eur. Heart J.* 12 (Suppl. E): 25-32.

Victor, R.G., L.A. Bertocci, S.L. Pryor, and R.L. Nunnally. 1988. Sympathetic nerve discharge is coupled to muscle cell pH during exercise in man. *J. Clin. Invest.* 82: 1301-1305.

Wallensten, R., and E. Eriksson. 1984. Intramuscular pressures in exercise-induced lower leg pain. *Int. J. Sports Med.* 5: 31-35.

Wallensten, R., and J. Karlsson. 1984a. Histochemical and metabolic changes in lower leg muscles in exercise-induced pain. *Int. J. Sports Med.* 5: 202-208.
———. 1984b. Local histochemical and metabolic changes in lower leg pain. In *The peripheral circulation*, ed. S. Hunyor, J. Ludbrook, J. Shaw, and M. Mcgrath, 87-95. Amsterdam: Excerpta Medica.

Warren, J.A., R.R. Jenkins, L. Packer, E.H. Witt, and R.B. Armstrong. 1992. Elevated muscle vitamin E does not attenuate eccentric exercise-induced muscle injury. *J. Appl. Physiol.* 72: 2168-2175.

Weber, P. 1989. Are we what we eat? *Proceedings of the International Conference on Fish Lipids and Their Influence on Human Health.* Svanoy, Norway: Svanoy Foundation.

Wefers, H., and H. Sies. 1988. Antioxidant effects of scorbate and glutathione in microsomal lipid peroxidation are dependent on vitamin E. *Adv. Biosci.* 76: 309-316.

Widmark, G. 1957. Autoxidation of (+)-limonene. *Arkiv Kemi.* 22: 211-213.
———. 1993. Autoxidation of methyl oleate (in Swedish). Apple Link [Online]. 15 March.

Williams, R.S. 1989. Contractile activity and expression of genes encoding proteins of oxidative metabolism in muscle. In *Advances in myochemistry.* Vol. 2, ed. G. Benzi. London: John Libbey.

Willis, R.A., K. Folkers, J.L. Tucker, L.-J. Xia, and H. Tamagawa. 1990. Lovastatin decreases coenzyme Q levels in rats. *Proc. Natl. Acad. Sci. U.S.A.* 87: 8928-8930.

Wilson, P.W.F., W.P. Castelli, and W.B. Kannel. 1987. Coronary risk prediction in adults (the Framingham Heart Study). *Am. J. Cardiol.* 59: 91G-94G.

Witt, E.H., A.Z. Reznick, C.A. Viguie, P.A. Starke-Reed, and L. Packer. 1992. Exercise, oxidative damage and effects of antioxidant manipulation. *J. Nutr.* 122: 766-773.

Wolbarsht, M.L., and I. Fridovich. 1989. Hyperoxia during reperfusion is a factor in reperfusion injury. *Free Radic. Biol. Med.* 6: 61-62.

Yamagami, T., N. Shibata, and K. Folkers. 1976. Bioenergetics in clinical medicine: Administration of coenzyme Q_{10} to patients with essential hypertension. *Res. Commun. Chem. Pathol. Pharmacol.* 14: 721-727.

Yamamoto, Y., E. Niki, Y. Kamiya, M. Miki, H. Tamai, and M. Mino. 1986. Free radical chain oxidation and hemolysis of erythrocytes by molecular oxygen and their inhibition by vitamin E. *J. Nutr. Sci. Vitaminol.* 32: 475-479.

Yu, C.-A., and L. Yu. 1981. Ubiquinone binding proteins in energy conserving systems. *Biochim. Biophys. Acta* 639: 99-128.

Zamora, R., F.J. Hidalgo, and A.L. Tappel. 1991. Comparative antioxidant effectiveness of dietary β-carotene, vitamin E, selenium and coenzyme Q_{10} in rat erythrocytes and plasma. *J. Nutr.* 121: 50-56.

Ziboh, V.A., K.A. Cohen, C.N. Ellis, C. Miller, T.A. Hamilston, K. Kragballe, C.R. Hydrick, and J.J. Voorhes. 1986. Effect of dietary supplementation of fish oil on neutrophil and epidermal fatty acids: Modulation of clinical course of psoriatic subjects. *Arch. Dermatol.* 122: 1277-1282.

Index

About the Author

Jan Karlsson, PhD, has conducted academic research on muscle metabolism for more than 30 years and on radicals and antioxidants for more than 15 years.

Karlsson served as a visting professor in the Institute for Biomedical Research at the University of Texas-Austin from 1981 to 1991. He was director of the Human Performance Laboratory at Karolinska Hospital in Stockholm, Sweden, from 1977 to 1982, and acting professor in sport exercise physiology at the Karolinska Institute from 1974 to 1977. Karlsson received his DrSci in 1971 and completed his post-doctoral work at Parkland Memorial Hospital in Dallas, Texas, in 1974.

Karlsson has written more than 250 articles and reviews about muscle physiology, molecular cardiology, exercise physiology, clinical cardiology, and drug testing, and he serves as a consultant to several major international pharmaceutical companies. He has received awards for his contributions in exercise medicine and drug development in Finland, the former West Germany, and Japan.

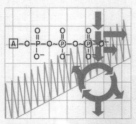
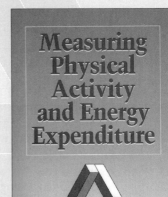